A Field Guide in Color to
MINERALS, ROCKS
and
PRECIOUS STONES

A Field Guide in Color to

MINERALS, ROCKS AND PRECIOUS STONES

By Dr. Jaroslav Bauer
Photographs by František Tvrz

OCTOPUS BOOKS

Translated by Zdenka Náglová
Graphic design by Antonín Chmel

English version first published 1974 by
Octopus Books Limited
59 Grosvenor Street, London W1

ISBN 0 7064 0305 3

Distributed in Australia by
Rigby Limited
30 North Terrace, Kent Town
Adelaide, South Australia 5067

Printed in Czechoslovakia by Svoboda, Prague
3/07/03/51

Contents

Preface

As early as prehistoric times men knew how to make use of the various special properties of individual stones, ores and minerals; in other words, Stone Age man was the first mineralogist. In the course of time the use and importance of minerals for technical and scientific purposes has been increasing and modern industry still finds them essential as raw materials. The rich colour, perfect lustre and the regularity of form of crystal have long had a particular charm for man.

This book of 576 colour photographs is sure to appeal not only to mineral collectors, but also to laymen who appreciate the beauty of attractive stones. The wondrous world of nature lies open here to all those who are interested in discovering its hidden beauties.

The photographic material was chosen from the taxonomic as well as the aesthetic point of view. The expert mineralogist will also find here unusual and rare specimens of minerals which he may have looked for in vain in other books dealing with the subject.

This book is not meant to be used just for the purpose of easy identification of minerals — that is not easy at all. As it is intended for the layman, the book does not presuppose that the reader will know much about crystallography, chemistry or physics. Because many minerals may display a number of different characteristic forms, it is rather difficult even for an expert to classify them. For this reason typical features have been chosen which can either be recognized on sight or with the help of very simple tests.

The most striking feature of a mineral is its colour. It is, of course, not always the most reliable deciding factor. In this book the minerals have been divided according to their colour. Whenever a certain mineral occurs naturally in more than one colour, it is especially mentioned each time. To make it easier for the reader to use the book, individual colours are given in the form of coloured strips along the margins of the tables and plates.

The second important feature of a mineral is its lustre. This book divides minerals into metallic and non-metallic. The identification tables arrange the minerals according to their hardness as well as in the general mineralogical system. To enable more accurate identification to be made some other easily discernible features, such as streak, cleavage and transparency, are described.

The majority of identification keys and books depict idealized crystal forms. In nature, however, such forms are very rare. The minerals actually seen in small or large collections mostly display different, distorted and completely irregular forms. Several specimens of a particular mineral, e.g. quartz, may vary not only in colour (see plates) but also in crystal form, e.g. the many crystal forms of calcite.

Perfect models of crystals are only confusing for beginners, and cause them many difficulties later when they expect to find them in the field. For this reason drawings of crystals have not been included in the tables, but there is a summary of the main crystal forms on page 15 to illustrate the terms used in column five of the tables. Readers who have a special interest in crystallography are referred to the list of reference books on page 208.

Precious stones and rocks are mentioned fairly briefly here. More detailed information may be found in numerous other books dealing with this special subject, and here too the reader is referred to the list of literature on page 208.

All the minerals described here form part of the excellent collection kept at the Technical University, Prague. Some of the specimens mentioned are from the Mineral Section of the National Museum, Prague.

Introduction

Basic Terms

Neither the apparently solid earth nor the porous soil under our feet are homogeneous. On the contrary, the upper layers of the earth's crust are composed of very different elements — in technical language, of heterogeneous building blocks. They are called *rocks*. At a closer glance it becomes evident that these are accumulations of minerals. *Minerals* must therefore be considered the basic constituents of rocks. In fact, rocks are generally mixtures of various sorts of minerals affected by natural processes. They form a heterogeneous yet geologically independent part of the earth's crust.

Although there are some rocks (such as marble, composed only of calcite crystals) which are chemically homogeneous but physically heterogeneous, other minerals can occur naturally which are chemically and physically homogeneous. Each one has a definite chemical composition. Apart from accidental inclusions and impurities, every mineral maintains the same character and composition regardless of where it is found.

Each mineral has special characteristic features and a typical and regular shape. Most are limited by plane faces, sharp edges and angles. Such minerals are called *crystals*. Their external regular shape is closely related to their internal structure. In other words, crystals are solids whose smallest parts, atoms, are arranged in a regular manner. The regularity of the external shape can be disturbed yet the orderliness and regularity of arrangement of the interior of the crystal remain unaltered.

Some minerals possess no geometrically regular shape. The internal organization of individual atoms is not always orderly, and the minerals are bounded by irregular faces. Such minerals are called *amorphous*, e.g. opals. Normally most amorphous minerals are easily subject to change and tend, sooner or later, to crystallize.

There are also the so-called *cryptocrystalline* minerals which outwardly resemble amorphous substances yet retain their inner crystalline structure, e.g. chalcedony.

All minerals are solid with one exception: mercury.

The majority of minerals occur in crystal form, yet not all minerals grow crystals.

Those minerals which contain elements important for industrial purposes and serve as raw materials for the extraction of these elements, are called *ores*.

Minerals

Material Structure of Minerals

The earth's crust is composed of various rocks whose separate constituents are minerals. Most are compounds of chemical elements. Some of them, however, are composed of a single element, such as copper, gold, sulphur, or carbon. The latter is better known in two forms as diamond and graphite. The basic element of the commonest mineral — quartz — is silica (SiO_2), a compound of silicon and oxygen. In many other minerals, on the other hand, the elements form complicated compounds. Most are of inorganic origin, but occasionally minerals with organic constituents are found, such as amber (fossil resin), or whewellite (calcium oxalate).

There are some 3,000 different minerals, composed of 92 chemical elements, from oxygen to uranium. Of the known minerals only a small number — some 40-50 — are generally rock-forming, e.g. quartz, felspar, mica, pyroxene, amphibole and olivine. These minerals form the major part of various rocks, and in consequence are called *rock-building minerals*.

System of Mineral Classification

As in all other branches of natural sciences, mineralogists have found it necessary and ufesul to arrange minerals in a system that is easy to understand, even though the figure of 3,000 known minerals is tiny in comparison with the tens of thousands of different plant and animal species in the world. It is convenient to arrange minerals according to their chemical composition and internal atomic lattice structure and almost all collections of minerals are arranged according to this system. It starts with the chemically simplest minerals, i.e. elements which are themselves minerals, and continues to more complicated compounds, while simultaneously considering the chemical relationship of individual minerals. This system is also recommended for arranging smaller private collections.

According to their crystallo-chemical properties minerals are usually grouped into nine natural classes corresponding to the mineralogical tables listed by H. Strunz:

1. Elements
2. Sulphides (selenides, tellurides, arsenides, antimonides, bismuthides)
3. Halides
4. Oxides, hydroxides
5. Nitrates, carbonates, borates
6. Sulphates (chromates, molybdates, wolframates)
7. Phosphates, arsenates, vanadates

8. Silicates
9. Organic substances

The Strunz classification system has been used to classify the minerals in this book.

Origin of Minerals

Magmatic Formation of Minerals

As the whole living world is in a continual state of change, so minerals also originate, grow and alter. Most of them are formed very deep down in the earth, where they are exposed to the effects of high pressures (thousands of atmospheres) and high temperatures (from approximately 900° to 1300°C). The earth's depths are composed of a glowing, liquid, molten silicate mass called *magma*. Because of the continuous movement of the earth's crust, parts of this magma have been driven upwards to cooler layers, where they gradually solidified forming agglomerations of rocks.

The magma is a molten mass, saturated with gases and water vapour and composed of different silicates and oxides. Its composition generally corresponds with the chemical properties of the rocks forming the earth's crust. Various currents keep the magma moving continuously, causing chemical reactions in its internal regions. Consequently new compounds arise, forming fresh minerals.

When the molten magma — which is constantly under pressure — reaches the upper, cooler levels of the earth's crust, its temperature falls. In the course of this cooling process the first minerals start to separate out, their number growing with the gradual cooling of the magma. The lighter minerals remain in the upper levels and the heavier ones sink slowly down again. This process is called *magmatic differentiation*. As a result, rich heavy mineral deposits are formed, such as those of magnetite and chromite.

During the next stage of crystallization the crystals start to grow. Minute cores become overgrown with crystals and the process continues as the magma cools.

In the final phase of crystallization the remainder of magma becomes more liquid. Its content of volatile components, such as gases and steam, is increasing. If some of the magma leaves the original mass it may form the so-called *pegmatites*, in which such minerals as mica, tourmaline and beryl are concentrated, together with others containing elements of rare earths, such as tin and tungsten ores. Finally the remainder of the magma cools off.

Part of the gases and vapour remains enclosed in the rocks and, like air-bubbles in a loaf, forms almond-shaped cavities. These cavities are a common phenomenon especially in basaltic rocks. At a later stage they often become filled with quartz, agate or chalcedony. The largest part of the gases and vapour escapes through rock fissures and cracks to the surface of the earth. Meanwhile the originally hot solution becomes cooler and new minerals appear in the form of crystals on the walls of the fissures. Well known minerals, such as quartz and calcite, originate at this stage and are known as *hydrothermal* minerals. If elements of heavy metals are present at this time ore veins are formed. The process of direct separation of specific ores, such as molybdenum, tungsten and tin ores, from hot gases and vapour, is called *pneumatolysis* or *pneumatolytic mineral formation*.

Just under the surface of the earth the steam changes into water but still contains many mineral substances. Together with surface water, which has seeped underground, it jets forth in the form of mineral springs. Cold as well as hot mineral springs give rise to various other minerals, such as aragonite or geyserite.

In places where hot solutions and gases escape through fissures and cracks in sedimentary rocks such as limestone, they dissolve the rocks giving rise to new secondary minerals. This is *metasomatic* mineral formation. Siderite-iron spar originates in this way.

Mineral Formation Caused by Weathering

On the surface of the earth all minerals and rocks are constantly subjected to the disturbing effects of various forces; this is called *weathering*. The effects of weathering are gradual, irrevocable and continuous. Rock surfaces are mechanically affected by changing temperature and by the shattering effects of frost. Chemical changes are caused by atmospheric oxygen, carbonic acid and water. Biological processes can also produce changes in rock surfaces.

The effects of weathering can cause profound changes in minerals; felspars change into kaolin, olivine into serpentine, and the golden-yellow pyrite into brown limonite. As a result of the weathering of pyrite sulphuric acid is liberated, which may affect the neighbouring calcite, changing it into gypsum or into a whole series of other sulphates. Opals are formed in a similar manner.

In this way several secondary minerals can originate from one mineral, e.g. malachite, azurite, or limonite from chalcopyrite. Dripstone caves are another product of weathering processes.

Mineral Formation Caused by Chemico-Sedimentary Processes

Various minerals are deposited from the sea water as a result of its evaporation, or of a change in its chemical composition. Rock salt, sylvine, gypsum, calcite and some iron ores, such as chamosite, also originate in this way.

Minerals and Rocks of Biological Origin

Apart from the effects of natural processes upon minerals and rocks, new minerals can be formed from mineral substances dissolved in water. Coral islands and limestone bodies are the product of living organisms. New minerals can also originate from the decomposed remnants of dead organisms, such as older or more recent deposits of phosphorite. Sulphur, salt-petre, pyrite and marcasite are also of biological origin.

Metamorphic Mineral Formation

The molten magma in the interior of the earth affects the individual layers of the crust by its temperature, pressure, and chemical reactions with various substances. This causes sedimentary rocks to change their appearance and their physical and chemical properties. This process, during which new secondary metamorphic rocks and minerals originate, is called *contact* metamorphism. Some sorts of mica, garnet and kyanite are formed in this way.

Summing up, it may be said that minerals originate under different conditions. In natural conditions a mineral seldom occurs by itself, almost always it is associated with several other minerals called the *accompanying* or *associated minerals*. Such a mineral assemblage is called *paragenesis*. Its formation is controlled by specific natural laws. A thorough knowledge of these helps in understanding the processes of mineral origin, and makes it possible to deduce from the occurrence of one mineral the probable existence of some other associated minerals.

External Form of Minerals

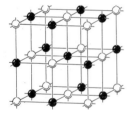

Halite (Rock salt) model of lattice

Crystals

In the preceding section we have briefly explained the terms crystalline and amorphous. We have also mentioned the internal structure of crystals which depends upon the arrangement of their smallest particles, ions, atoms and molecules. They are arranged in a geometrical pattern forming the *crystal lattice* or the *space lattice*. Several types of lattices may be distinguished according to the arrangement of the crystal matter. Some minerals such as NaCl (rock salt) display comparatively simple space lattices, whereas in other minerals very complicated space lattices can be observed. The grouping of atoms inside a crystal is expressed by the arrangement of its external faces. As a geometrical body the crystal is bounded by faces, edges and angles of different size and form. Their mutual position and symmetry follow the same natural laws as the internal structure.

Every type of crystallized mineral shows typical angles at which individual faces intersect. These angles — called *space angles* — are identical in all crystals

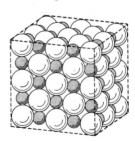

Internal structure of the crystal (grey balls — sodium, white balls — chlorine)

13

of one mineral irrespective of their size. This quality is defined by the Law of Constancy of Angle. It holds true, of course, only for identical adjoining faces.

Crystal Symmetry

One of the most striking features of crystals is the regularity of arrangement of the faces in juxtaposition with one another. This is called *crystal symmetry*. Listed below are the elements of symmetry:

1. The *plane of symmetry* divides the crystal into two symmetrical halves. For this reason it is often called the *mirror (reflection) plane*.
2. The *axis of symmetry* is an imaginary line passing through the centre of the crystal. Rotated about this axis through 360° the crystal returns to its original position. The axes of symmetry are divided into diad, triad, tetrad and hexad axes according to the number of times an identical face is displayed during rotation through 360°, i.e. a crystal which shows two identical faces during rotation is a diad, and so on.
3. The *centre of symmetry*: Each face of a crystal corresponds to a parallel, opposite face which is rotated through 180° about this imaginary centre.

There are in all 32 *crystal classes* in which the elements of symmetry are arranged according to their number and kind. They are grouped into seven major divisions, the seven *Crystal Systems*. Common characteristic features of individual systems are the so-called *crystallographic axes*. As three-dimensional systems of coordinates they enable a precise determination of the location of every crystal face. The smallest number of symmetry elements is found in the triclinic system, followed in a progressive sequence by the monoclinic, orthorhombic, tetragonal, trigonal, hexagonal and cubic systems.

Some representatives of the large number of different crystal forms of the various systems are shown on the opposite page.

Most minerals belong to a single class and consequently also have a certain definite type of lattice. Some minerals of identical chemical composition, however, having originated under quite different conditions, can crystallize in two or more classes. Such minerals are called *polymorphs*. The individual crystal types in which polymorphic minerals occur are called *modifications*, e.g. diamond — graphite, calcite — aragonite, pyrite — marcasite, quartz — cristobalite — tridymite, rutile — brookite — anatase.

Appearance and Development of Crystals

The assemblage of all faces of a crystal is called the *form*. The relative width and length of these faces determine its shape which is called the *habit*. These may range from a pyramidal habit, in which all corners converge through columnar to tabular, acicular (needle-like), fibrous, tabular, plate-like or lamellar to the isometric habit with all faces equally developed.

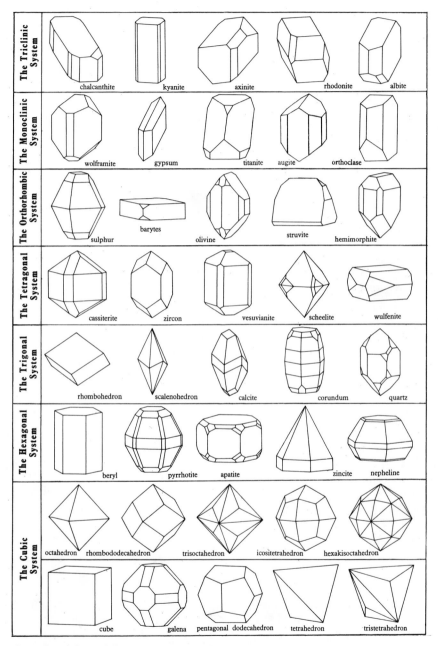

Examples of Crystal Forms of the Seven Crystal Systems

15

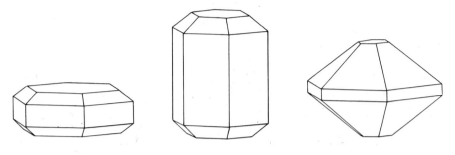

Different habits of crystals with identical combinations of faces

Two crystals of identical form may have different habits and vice versa. The form as well as the habit of a crystal depend upon its chemical composition, but both of these can be affected by physical conditions prevailing at the time of crystal's origin, such as temperature, pressure, speed of growth and character of the environment. The drawing above shows the different habits in crystals of identical form.

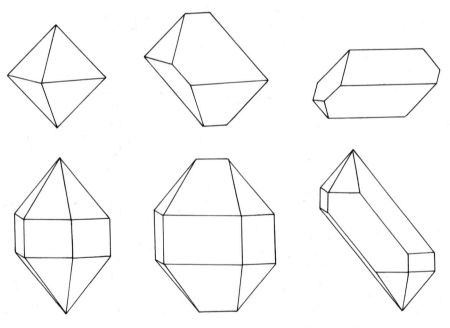

Crystal 'irregularities'. Left: ideal form, right: two distorted forms. Top row: magnetite crystals. Bottom row: quartz crystals

Crystals with even faces do not often occur naturally. However, they can originate in a soft, yielding environment, such as crystals of gypsum in soft clay. Crystals may also develop freely on the walls of caves. Such crystals always have well developed end faces. On the other hand, a one-way or irregular supply of new material in the course of the growth of a crystal can result in the most varied *distortions*. Such strangely developed forms can differ considerably from the ideal crystal forms as shown in the drawing on the preceding page. Nevertheless these 'irregularities' display angles of a constant value between similar faces.

Aggregates

Very often two individual crystals growing in a confined space have little room for the development of a new crystal. The result is an assemblage of interlocking crystals in irregular clumps called an *aggregate*. In rock crevices groups of crystals called *druses* or *geodes* grow from a common base in oval cavities. The appearance of these aggregates may differ according to their structure and morphology. On the fine internal surfaces of some limestones, skeletal and dendritic forms resembling fossil impressions of plants can sometimes be found. These so-called *dendrites* are manganese and iron oxides and hydroxides, and are entirely chemical in origin. They have nothing in common with fossils.

Growth of Twin Crystals

A characteristic feature of the growth of crystals in some minerals is the growth of *twin crystals* which is regulated by certain definite laws. In twin crystals re-entrant angles (interior angles greater than 180°) occur quite frequently, e.g. in cassiterite gypsum, and fluorite.

In some cases the growth of twin crystals results in a *pseudo-symmetry*, i.e. crystals of lower symmetry are twinned in such a way that they form a group of crystals of higher symmetry. For example, the orthorhombic aragonite often

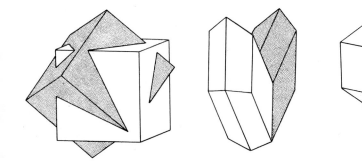

Twinned Crystals of (from left to right): fluorite, gypsum, cassiterite

grows columnar crystals of pseudo-hexagonal symmetry. The surface of such crystals is finely striated. This is called the *twinning striation*. Regular twinning of crystals in different ways is called *syntaxis* or *epitaxis*.

Crystal Faces

As mentioned before, crystals rarely have smooth even faces. More often their faces are considerably deformed and striated. Some minerals, such as pyrite, quartz, and tourmaline display such typical striations that they may serve as an important clue for identification.

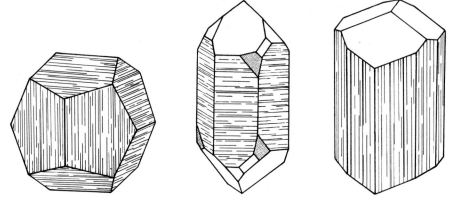

Characteristic striation of (from left to right): pyrite, quartz, and tourmaline

Pseudomorphs

Pseudomorphs are unusual mineral formations. They are crystals which have borrowed their form from another type of crystal, e.g. externally they may resemble the pyrite crystal yet their composition corresponds to that of limonite. This is the result of a partial or complete exchange of chemical constituents.

Physical Properties of Minerals

The internal arrangement of atoms and their force of cohesion determines not only the external form and the symmetry of crystals, but a whole series of other important physical properties of minerals. According to the character of the particular structure, some physical properties in a certain mineral may vary according to the direction from which they are handled. Mica, for instance, can be split only in one direction. Consequently, we may say that crystals are *anisotropic* bodies, i.e. they show different physical properties in different directions. In amorphous materials, on the other hand, all physical properties are independent of the direction from which they are affected. Therefore amorphous materials are *isotropic*, i.e. they behave in the same way in all directions.

Every mineral has some typical physical properties. These are an important aid in identification.

Specific Gravity

The specific gravity of a mineral is the relative number indicating how many times heavier or lighter it is than the same amount of water. The exact determination of specific gravity requires special methods. The approximate specific gravity may be estimated by simply balancing the mineral in the hand and guessing its weight. Most minerals show the specific gravity of 2 to 4.

A little practice in the estimation of the weight of pieces of different minerals of similar size may help in acquiring the ability to distinguish between light minerals (specific gravity 1—2), medium-heavy minerals (specific gravity 2—4), heavy minerals (specific gravity 4—6), and very heavy minerals (specific gravity more than 6).

Properties of Cohesion

Other important properties of minerals depend upon the structure of their crystals and the direction in which the crystals are affected. These cohesion properties of crystals are called hardness, cleavage, fragility and tenacity (ability to resist damage).

Hardness

A crystal of calcite can be scratched quite easily with a knife. If we try the same with a quartz crystal the knife slides over its surface. The two minerals have a different hardness, or better still, scratch hardness. The term *scratch hardness* of a crystal face indicates its power of resistence to mechanical injuries.

For the determination of the hardness of minerals a *hardness scale* was established by Friedrich Mohs (1773—1839). It has ten grades in which each higher grade of mineral can always scratch the one just below it in hardness.

The set of standard minerals used as examples for the hardness tests is as follows:

1.	talc	6.	orthoclase felspar
2.	gypsum	7.	quartz
3.	calcite	8.	topaz
4.	fluorite	9.	corundum
5.	apatite	10.	diamond

These hardness tests can be applied only to fresh, unweathered minerals.

The hardness of a mineral can be determined quite easily. Minerals of the first grade can be scratched by the finger nail, and leave a soft greasy feel in the hand. Minerals of the second grade can also be damaged by the finger nail. A piece of copper wire or a copper coin (hardness approx. 3) scratches minerals up to the hardness of 3. A pocket knife (hardness approx. 6) scratches minerals up to the hardness of 5. A good file scratches even quartz. Minerals of a hardness exceeding 6 can scratch window panes (hardness approx. 5). On the other hand minerals of grade 8—10 cannot be scratched by a file, and can be used in making sparks.

When determining the hardness of a mineral we proceed upwards from minerals of the lowest grades of the scale as long as we succeed in scratching them, each time trying the grade above for checking purposes. In the case of minerals embedded in rocks we proceed in the opposite way, i.e. we scratch the mineral to be tested with individual standard samples from the scale starting from the highest grade. After some practice it is quite easy to say whether the mineral scratches or only slides over the surface of another mineral. Usually the scratching is accompanied by a slight crackling sound.

Experience has shown that

1. the hardness of the mineral to be determined and the sample from the scale is identical if they do not damage each other;

2. the hardness of the mineral under determination and the scale sample is the same if they scratch each other. The angles and edges of a crystal are sometimes harder than its internal crystal or cleavage planes. Consequently, the edge of a gypsum crystal can scratch the cleavage plane of another gypsum crystal.

3. if the mineral under determination cannot scratch the mineral sample, but can itself be cut by it, it has a hardness somewhere between that of the sample and that of the next lower grade. In this case we consider the hardness to be of the lower grade adding a half-grade to it.

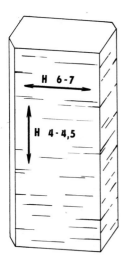

Different scratch hardness on kyanite

Although the determination of the hardness may seem to be quite easy, there are a number of factors which may be misleading. Thus the hardness of the same mineral may vary in different directions. It shows most clearly in the columnar kyanite. In the vertical direction kyanite shows the hardness 4-4.5 — it can be cut by a knife. Horizontally, however, it shows the hardness 6-7, and cannot be scratched by a knife. Scratch tests, therefore, must always be tried in different directions.

Sometimes soft aggregates are intergrown with some harder minerals, which make the hardness of the aggregate seem to be higher than it actually is. In such cases the hardness may show large differences, and a false identification may lead to completely wrong conclusions.

Other mistakes may arise if a piece of finely

granular and thick aggregate is harder than the individual crystals of the same mineral. For instance, a compact piece of gypsum can be scratched by a finger nail only with difficulty. On the other hand, fibrous and porous aggregates often seem to have a lower hardness because of the blank spaces between individual grains. Chalk may easily be cut by a finger nail in spite of the fact that it has the hardness 3. Weathered pieces of minerals behave in a similar manner.

In earthy minerals, efflorescences, needle-shaped, fibrous and fine-grained aggregates, the exact determination of hardness by such simple methods is quite impossible. In these cases it is advisable to make use of some other characteristic identification feature of the mineral.

Cleavage

When hit with a hammer or pressed with a knife some crystallized minerals break apart along surfaces related to the internal crystal structure, resulting in smaller pieces of minerals with new smooth cleavage planes. Minerals tend to cleave along planes of least cohesion, and some may split in several directions, i.e. rock salt and galena into cubes, fluorspar into octahedra, and calcite into rhombohedral fragments. Other minerals, such as mica and gypsum possess a perfect cleavage only in one direction, in other directions cleavage being quite imperfect or completely absent.

The cleavage of transparent crystals is often indicated by the delicate cleavage rifts — apparent on closer inspection — running along the crystallographic planes. (See drawing)

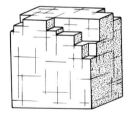 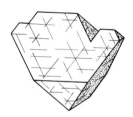 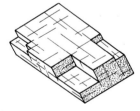

Cleavage (left to right): galena, fluorite and calcite

Fracture

Some minerals, such as quartz and opal, do not cleave in any particular direction, but fracture irregularly instead. The appearance of the fractured surface may be even, uneven, conchoidal, subconchoidal or hackly (jagged). Metals and tough minerals have a hackly fracture. The type of fracture can also be used to distinguish between mineral species.

Other Physical Properties Dependent upon Cohesion

Brittle minerals such as pyrite, quartz and opal, crumble into small fragment with a crackling sound when hit with a hammer. *Soft* minerals like talc or graphite split into powder under a hammer blow.

Ductile or *malleable* minerals, such as native metals or argentite, can be flattened out into thin plates and no powder appears when they are scratched.

A thin plate of mica can be bent by the fingers. If the pressure is removed it returns to its original position. Such minerals are called *elastic*. Others, such as gypsum or antimonite, can be bent but they do not spring back to their original position. They are called *flexible*. On the basis of these properties we can easily distinguish between certain similar minerals, e.g. the elastic mica from flexible chlorite.

Colour

Some crystals have especially clear, brilliantly shining colours. It is no wonder that various colour shades have been named after them. We are familiar with emerald-green, ruby-red, turquoise or azure blue, amethyst-purple, and so on.

The colour of a mineral is often its most striking property, playing an important part in its identification. Yet it is not always a reliable clue or a characteristic feature of a particular mineral. There are several *idiochromatic* minerals which have their own and unchangeable colour, malachite is always green, graphite always black, and sulphur always yellow. In these minerals colour is a typical feature. There are, however, many other minerals which in their pure form are *colourless*, such as rock crystal (quartz), calcite, and rock salt. Yet if they contain any impurities they may be of quite different colours, such as blue salt, and yellow, pink, violet or brown quartz. Fluorspar can exist in almost all colour shades. Tourmaline, apatite and beryl also occur in many colours. The variable colours of these *allochromatic* minerals cannot be used for identification purposes as their colouring can be due to different impurities, pigments, mineral inclusions, and occasionally to small traces of other elements. The colouring may also be affected by radioactivity.

Some minerals change their colour with a change of illumination, e.g. alexandrite is green in sunlight whereas in artificial light it is violet. There are also minerals whose crystals change the intensity of their colour when held up to the light and turned around. For instance, when rotating a crystal of cordierite its colour changes from blue to yellow. Such minerals are called *pleochroic*. The change of colour is caused by the crystal's unequal absorption of light vibrating in different planes.

The colour of a mineral may sometimes be concealed under an encrustation. Numerous ores *tarnish* on the surface. Therefore their colour must always be tested on a fresh surface. The test is best carried out in a diffused light rather than in sunlight or artificial light.

22

The above-mentioned alexandrite changes its colour temporarily if exposed to the effects of light. Numerous precious stones lose their brilliant colours on continued exposure to the sun's rays. The emerald loses its rich green colour, and the amethyst and the rose quartz become paler in sunlight. Minerals containing silver also react sensitively to light, e.g. pyrargyrite and proustite. When exposed to the light they slowly become covered by a thin black film. In museum collections such minerals must be protected from the light. The deep-red realgar reacts in a similar way when it changes into the yellow orpiment in the presence of light. All these *changes of colour* take place only slowly.

A long time ago men learned how to speed up the natural processes of colour change. Yellow citrine or gold brown topaz may be obtained from the violet amethyst by heating. Diamonds, rubies and sapphires display richer colour shades under radioactive or ultraviolet rays. Radiation changes rock-crystal quartz into smoky quartz. Agates which are of an indistinct grey colour can be 'coloured' artificially by boiling in an aniline dye, in much the same way as textile fabrics are dyed.

Streak

Fortunately one characteristic — the *streak* of a mineral — is constant irrespective of colour variations. A mineral to be identified is rubbed on a piece of roughened or unglazed porcelain called a *streak plate*. It leaves behind a coloured scratch which is an important identification sign. We must not forget, however, that the hardness of the porcelain according to Mohs' scale is 6—6.5. Therefore minerals harder than this will produce only white porcelain powder. If we intend to determine the streak of a harder mineral we must either rub it to powder in a mortar or crush it with a hammer on a small anvil. The powder can then be rubbed on the streak plate. In this case only fresh unweathered minerals should be used.

Coloured minerals always produce a coloured streak which is a little lighter than the true colour of the mineral. Colourless and white minerals (of non-metallic lustre) need not be tested since their streak is always white. Similarly, discoloured minerals generally produce a white or light-grey streak. In this case the streak is likely to be discoloured as well. Its colour is indistinct since its intensity depends upon the proportion of pigment and other impurities. The greatest difference between the true colour and the streak of a mineral is shown by minerals of metallic lustre. Yellow pyrite produces a greenish-black streak, black haematite gives a red powder, black wolframite has a brown streak, and black cassiterite an almost colourless streak. Test results have shown that the determination of the streak is an important and simple means of identification. Naturally minerals producing a coloured streak are easier to identify than those producing a colourless or discoloured streak.

23

Lustre

Apart from the colour the *lustre* is also an important aid in the identification of minerals. It generally depends upon the amount and type of reflection and refraction of light as well as upon the quality of the mineral surfaces. There are minerals of *metallic* and of *non-metallic lustre*. A particular mineral does not always have the same lustre. It is therefore not always feasible to distinguish between metallic and non-metallic minerals. Some authors have introduced a supplementary term *submetallic lustre* for minerals which lack the full lustre of the metals.

Opaque ore minerals, such as pyrite, galena, etc., display a metallic lustre, and possess a good refractive capacity. With another large group of minerals, such as the dark sphalerite, cassiterite, rutile, etc., it is rather difficult to specify the type of lustre, therefore they appear several times in the accompanying tables. The lustre of non-metallic minerals may be subdivided, according to intensity and quality, as follows: 1. *Adamantine* or diamond lustre, occurs in transparent minerals of good refractive capacity; 2. *Vitreous*; 3. *Greasy*; 4. *Dull*, in minerals of low refractive capacity. Special kinds of lustre are either due to the structure of the aggregate or to a perfect cleavage. *Silky* lustre is often found in fibrous aggregates, such as asbestos. Some minerals composed of fine layers have a *pearly lustre*.

Transparency

The *transparency* of minerals varies greatly. A mineral is *transparent* when writing can be read through it. Rock crystal (quartz), rock salt and topaz are good examples. Impurities and inclusions can cause transparent minerals to lose some of their transparency and to become *translucent*. Minerals of metallic or of dull lustre are called *opaque*.

The structure of a mineral determines its degree of transparency. Finely granular aggregates, such as gypsum and mica, are opaque or merely translucent. But larger crystals of the same minerals are perfectly transparent. Under magnification thin sections of finely granular aggregates also become transparent.

Other physical properties of minerals can be mentioned only very briefly in this book. Those who would be interested in more detail are referred to the more comprehensive literature.

Refraction of Light

The *refraction of light* is an important optical property which remains constant for each mineral. Its exact determination requires special instruments called refractometers.

Double refraction occurs when a ray of light passing through an anisotropic crystalline substance is split into two. A piece of writing placed under such a mineral is seen twice. *Double refraction* is shown especially well by the colourless, transparent crystals of calcite.

24

Luminescence

Exposure in the dark to ultraviolet light elicits coloured luminescence from some minerals, such as scheelite and willemite. Others go on emitting light after having been subjected to certain conditions, such as heating, e.g. fluorite. This ability is called *thermoluminescence*. Another kind of luminescence, *triboluminescence* can be induced by crushing or rubbing some minerals.

Luminescence is particularly characteristic of certain minerals. It may therefore be applied in mineral identification as another simple supplementary factor.

Heat Conductivity

Taking a piece of copper in one hand, and a piece of amber in the other, we can feel that copper is colder. This is due to the different heat conductivity of the two substances. An experienced stone polisher is able to distinguish real gemstones from glass imitations in this way, feeling the stone against his cheek, since the skin on the cheeks is especially sensitive to warmth.

A similar category includes such *physiological properties* as the feel. Graphite and talc, for instance, feel greasy and smooth in the hand; chalk and kaolin feel dry and rough.

Minerals which are soluble in water, such as salt, sylvite, epsomite, etc., have a characteristic taste, being respectively saline, bitter, and sour.

Other minerals, such as sulphur, arsenopyrite, and fluorite can be identified by the characteristic odour especially when the minerals are struck or rubbed.

Magnetism

Pieces or powdered fragments of some minerals are attracted by a magnet, particularly when they contain a large amount of iron. In this way they can easily be distinguished from other minerals. Magnetite and pyrrhotite being themselves magnetic, attract iron dust. Other minerals, such as haematite, become magnetic as a result of heating.

Chemical Properties of Minerals

The identification of minerals by their chemical composition requires some special equipment and a thorough knowledge of analytical chemistry.

There is a simple chemical process which can be tried even by a layman; this is to remove carbonates from other similar minerals by means of dilute hydrochloric acid. By using a drop of this acid, white calcite can be distinguished from a similarly white piece of gypsum. The acid remains unchanged on the surface of gypsum yet it dissolves the calcite. Carbon dioxide is given off during this process, i.e. the drop changes into foam.

Identification of Minerals

The preceding sections have shown that it is not easy to identify positively a mineral at first sight. A reliable identification can be made either by using quantitative analysis, or by determining the internal crystal structure. For this a specially equipped laboratory is needed as well as the corresponding special knowledge.

Nevertheless a collector and admirer of minerals can identify many of them without any chemical analyses, from their external shape and physical properties. There are several identification clues which can be applied, based upon some typical mineral features. They include the morphological and the optical properties (refraction index, double refraction), the chemical composition (simple reaction tests), and the physical properties. Another way of identification is through the gradual elimination of the mineral species being considered in their class order. Common diagnostic features can only be used with typical mineral specimens. In other cases where it is necessary to employ all possible means of identification, or where only a few characteristics are really typical and reliable, their application would be too complicated. Amateur collectors are advised to pay attention only to those properties which may be recognized at first sight or by some very simple test.

Process of Mineral Identification

As a starting point for mineral identification, use a piece of fresh mineral broken off from a larger specimen. Testing should be very thorough and all properties recognizable by the naked eye or by means of a lens should be carefully recorded. Such a record may prove useful in cases when no explicit result has been achieved, and the identification of some properties may have to be repeated.

First note the external shape and appearance of the mineral, and in polycrystalline minerals note also the arrangement of crystals, the twinning and the symmetry. Determine the colour of the mineral and make sure by observing the streak that the mineral has not merely been tinged. The definition of the lustre may often seem rather complicated. It is therefore advisable to note down several different possibilities. The same applies to transparency. Hardness should be determined at first only approximately, more precise determination being possible at a later stage by means of the Mohs' Scale of Hardness. The more precisely the degree of hardness is determined, the more the search can be narrowed down to the correct place in the tables, to which the material under test is likely to belong.

Next, observe the presence or absence of cleavage, examine the cleavage planes where present, then test for brittleness, malleability, flexibility and elasticity. By weighing a mineral specimen in the hand we can estimate its specific gravity. With a smaller fragment of the mineral its solubility in water or hydrochloric acid can be tested. If the test specimen dissolves with effervescence in hydrochloric acid, it belongs to the carbonate class. Note also the paragenesis, i.e. the associated minerals, as these may be a guide to the identification of the specimen. On the basis of the observed properties we find the number of the mineral in the index on pp. 188—197, and under the same number in the identification tables a more detailed description of its properties will be found. To aid in identifying minerals existing in several colours, lustres and transparencies, they are included in the tables in several places.

Having finished the identification of the mineral check the results once more, comparing them with the data in the book. Ideally they should be identical. In practice the following situations may occur:

1. The mineral can be identified at first sight, but check the details once more to be certain.
2. Several minerals are very similar in their external shape. In successive tests examine their properties carefully and eliminate those that do not fit the description.
3. The mineral is completely unknown. In this case all its properties must be checked and eliminated in succession until a definite identification can be made. The more properties recognized, the easier the identification becomes.
4. It is impossible to identify the mineral with the aid of the tables. In such a case a mistake has probably been made somewhere. More attention should have been paid to similar minerals mentioned in the identification tables. Some seemingly unimportant feature may have been overlooked, or the mineral — whose identification might require some more complicated procedure — is not included in the tables at all.

Examples of Mineral Identification

Let us start from the fact that by means of the external features, such as lustre and colour, and with the aid of the tables, a preliminary conclusion has been reached according to which the mineral under investigation may belong to two or more mineral classes. The following section will show some examples of how to proceed in such a case, and which features should be selected to make identification easier.

The granular aggregates of magnetite and chromite look very similar. However they can be easily distinguished by the colour of their streak, which is black in magnetite and brown in chromite. Apart from this, magnetite is much more strongly magnetic than chromite.

If no crystal boundary is present, compact pyrite showing an iridescent surface, and chalcopyrite, are similar in appearance, the colour of their streak also being identical. They can be distinguished only by the degree of their hardness, i.e. pyrite is harder (6) than chalcopyrite (4.5).

Black sphalerite and black cassiterite display the same submetallic lustre. In this case also, they can be distinguished by their degree of hardness. The hardness of sphalerite is 3.5, and of cassiterite 6.5. A supplementary distinctive feature may be the frequent association of sphalerite with other sulphides, and of cassiterite with quartz and mica.

Chalk and kaolin can easily be distinguished chemically by using dilute hydrochloric acid. Chalk, as calcium carbonate, dissolves in it with effervescence.

Fluorite, amethyst and apatite can be recognized by the form of their crystals. However their coarsely granular aggregates are all of a similar violet colour. Amethyst is the hardest of the three minerals; it cannot be scratched by a knife, and is not cleavable. Fluorite shows a perfect cleavage, and when heated it becomes luminescent. Apatite is one degree harder than fluorite, and shows a different cleavage.

Augite, amphibole and tourmaline occur frequently in the form of black columnar crystals. Tourmaline may be identified by its poor cleavage and the typical striations on its prism faces. It occurs predominantly in acid igneous rocks. On the other hand, augite and amphibole often occur in basic igneous rocks. They can be distinguished from one another by the angle and quality of their cleavage planes. Augite has good cleavage at 90° angle, amphibole perfect cleavage planes at 55° and 125° angles.

Felspar, calcite, barytes and gypsum are often of a whitish colour, and all of them show a similarly perfect cleavage. Gypsum can be eliminated since it can be scratched by a finger nail, and felspar because it can be scratched by a knife. Barytes and calcite are of an identical hardness and differ only in weight, i.e. barytes is heavier. Calcite, on the other hand, readily dissolves with effervescence in dilute hydrochloric acid.

The few above-mentioned examples cannot, of course, fully illustrate all the problems of mineral identification. Nobody should be discouraged by initial bad luck, it is sometimes difficult even for an expert to identify some of the rarer minerals. In case of doubt every serious collector is sure to find assistance and advice from some experienced mineralogist. The best help, however, is our own perseverance and endurance.

Practical Advice for Collectors

Everyone starting a collection of minerals should determine in the first place what he is going to collect, and how his collection is to be set up. It is easiest to start first by collecting minerals and rocks occurring locally. Since the mineral richness of individual areas varies considerably, so will the extent of one's first collection. In any case the collector is advised to begin by making himself thoroughly acquainted with the geological conditions of his future collecting area. A good geological map, some special literature and the advice of other experienced collectors will all be helpful. Having this preliminary information the collector is ready to start a systematic collection of all mineral specimens which are typical of his locality.

Most probably in the initial stages each collector will pick up every possible mineral specimen within his reach. In the course of time, however, he is likely to replace some of the original pieces with better specimens, and he will be more aware of the aesthetic aspects of his collection. The value of a well organized and arranged mineral collection lies not only in the pleasure derived by the collector. It may serve later as research material for professional mineralogists. Frequently, an amateur collector possesses better specimens than a state museum.

Little by little one's collecting area can be enlarged and the collection expanded through exchange or purchase until one finally possesses a so-called systematic collection including samples of all attainable mineral types. Apart from the fact that a mineral collection is never complete it takes up much space in the home. He who has only a limited space for his hobby can establish a 'mini-collection' including only small but carefully selected and perfect specimens. An attractive collection can be made up of small individual crystals of only a few minerals, yet each displaying a different type of habit. A collection of important rock-building minerals or rocks in their most varied forms can also be interesting as well as instructive.

Where are Minerals to be Found?

Minerals cannot be planted like flowers. They are gained painstakingly after a long, patient search. Yet, it is this seeking and finding that brings the greatest pleasure to collectors. Where can they most easily be found?

First of all, any local stone quarries should be investigated very thoroughly as soon as possible after a blast. Do not forget, however, to ask the right authorities for permission! Abandoned quarries frequently yield a rich crop. The danger of falling stones should not be underestimated in the excitement of the moment. One should always proceed very cautiously, and carefully.

A large number of interesting minerals can also be found on heaps of mine refuse, but never enter abandoned shafts or caves alone. This is very dangerous even for an experienced collector.

A good opportunity for finding minerals and rocks arises, for instance, during the construction of roads, retaining dams and tunnels. In the course of sinking a well or digging a grave the changing composition of rocks with the increase in depth is clearly seen. Minerals can also be collected on the rocky banks of rivers and brooks, in rocky river beds or gravel pits. Interesting mineral specimens are also found on arable land or in stone debris along the roadside.

Collecting Equipment

The most useful tool in mineral collecting is a good geological hammer made of hard steel. There are pointed hammers and hammers with sharp edges diagonal

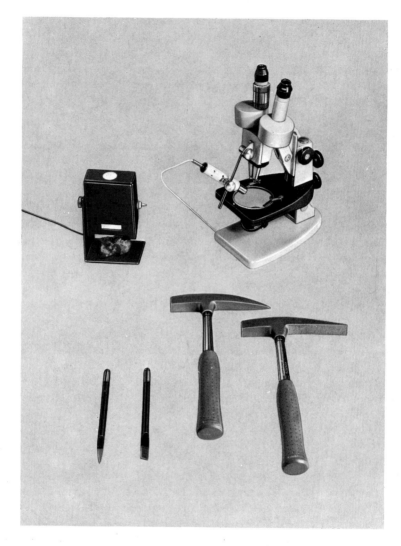

Equipment for mineral collectors. Above: UV-analyzation lamp for testing minerals by ultraviolet light. Right: stereoscopic prism lens for mineral identification and preparatory work. Below: pointed and flat chisels, geological hammers with pointed and with cutting edges.

to the handle (see illustration). The handle is usually 30 cm long, its approximate weight being 900 gm. To crush larger pieces of minerals we use a heavier, short hammer. For removing the superfluous remnants of rock and for the final trimming of the sample a small hammer (200 — 400 gm) similar to the geological hammer is normally used. The lighter the hammer, the longer the handle must be to give strength to the impact. As fundamental equipment, however, a single geological hammer will do.

Apart from the hammer, different chisels are needed (see illustration), i.e. pointed as well as flat. They should be made of good, hardened material. The ideal length is 20 cm. By means of the chisels specimens of minerals can be prized out of cracks and crevices without damage. They can also be used for breaking open quartz geodes.

Specimens are usually carried in a knapsack. It pays to buy a knapsack of a good quality with a supporting frame, since minerals are hard and very heavy. For wrapping the specimens to prevent them damaging each other, one can use newspaper, soft paper or cotton, the most brittle ones being placed in cardboard boxes. Each specimen should be provided with a small label (4 × 6 cm) indicating where it was found. Numbered labels, adhesive tapes, a notebook, pencils and some cards should always be at hand.

An indispensable aid in field work as well as at home is a lens, magnifying at least eight- to ten times. The exact identification of a mineral in situ is almost impossible, so it has to be taken home to be examined further. The following implements are essential: a lens, a hammer and some chisels, pincers for breaking off smaller pieces, an iron plate, and if possible a small grinding mortar. Other indispensable aids are: small streak plate of rough hard porcelain of about the size of a match box; a Mohs' Scale of Hardness (at least for the first 8 grades); a piece of broad glass; a knife with a steel tip; a magnet or magnetic needle; a steel needle; a piece of copper wire or a copper coin; a good file and a pair of forceps; a hermetically sealed bottle with dilute hydrochloric acid and a drop pipette.

Field Work

The same care that was used in collecting and treating the minerals must also be used in taking notes of all observed details before the minerals are packed for transport. Use plenty of paper — it is advisable to wrap every individual specimen carefully. Each must be provided with a label (or numbered tag) indicating where it was found, and then wrapped in another piece of paper. Never place the label directly on the mineral, it could be damaged during transport. The wrapping should be as tight as possible to prevent the specimens from moving and tearing the paper cover. Finally close the package by means of adhesive tape. Exposed crystal surfaces should be covered first with a layer of crumpled soft paper or a cotton plug, and then wrapped up in newspapers. Delicate mineral specimens should be embedded in cotton-wool and placed in cardboard boxes. The finest and most delicate minerals, such as the needle-shaped natrolite crystals cannot even stand the touch of cotton-wool. Such specimens are transported fastened with a piece of thread or a fine wire to the bottom of a cardboard box.

The number of individual labels (or tags) indicating the original locality should agree with the number in the notebook. Apart from the number the label should bear the date, the locality and the presumed mineral type. The description of the locality should preferably be more detailed, e.g. walls of a quarry, or refuse heap specimen, so that one could find the place again if need be. Moreover a mineral loses its value if its locality is unknown.

Preserving and Labelling Collected Specimens

After carefully unpacking the specimens at home they must be dusted and cleaned. Less delicate and larger pieces can be washed with water, soap and brush. Finer specimens are better cleaned with a fine brush. From needle-shaped crystals the dust can be simply blown off or they may be carefully rinsed in lukewarm water. Minerals which are soluble in water, such as salt, should be washed with petrol. Some sulphides, such as pyrite and marcasite, should not be treated with water, as this may cause them to decompose later. They are best cleaned with spirit. Crystals of gypsum are sensitive to soap which covers them with a whitish film. Minerals should never be exposed to direct sunlight to dry.

In order to mark a specimen a small label with a number can be affixed to the reverse side where it is least obtrusive. Or a small face of the mineral can be painted with white enamel and the number written in black Indian ink upon it.

Very small crystals are placed in glass tubes with their number fastened to the cork. Similarly, minerals soluble in water are kept in glass tubes which are sealed with wax or paraffin.

For an easily surveyed collection of minerals and rocks, special collection boxes with labels for marking the minerals are available. Every label should bear in its upper corner the name and the address of the collector, the number, and the date it was found. The middle part is reserved for the name of the mineral (in large letters) followed below by a short description (associated minerals, weight, size). Below left the locality is given, below right the manner in which it has been obtained.

A serious collector compiles a list — a catalogue — of his collection. The catalogue should contain the data on the labels and can be made more complete by including a more detailed description of the locality, references to special literature, conditions of exchange, and purchase or price. With larger collections it is also advisable to set up a card index of minerals in alphabetical order.

Every collection of minerals must be carefully protected from dust and humidity; excessive dryness and great variations in light and temperature can also be harmful.

Notes to the Identification Tables

The colour sequence of minerals: 1. colourless and white, 2. yellow, 3. orange, 4. red, 5. violet, 6. blue, 7. green, 8. brown, 9. grey, 10. black, 11. multicoloured.

In the identification tables the following terms are used for individual crystal systems: triclinic, monoclinic, orthorhombic, trigonal, tetragonal, hexagonal, cubic and amorphous minerals. Within the scope of individual colours, minerals of metallic lustre are indicated first, followed by those of non-metallic lustre. Within these two groups minerals are arranged according to their hardness from H 1 to H 10.

Abbreviations

assoc.	—	associated
chem.	—	chemical
deps.	—	deposits
diff.	—	different
nat.	—	native
No.	—	number
o.	—	other
phys.	—	physical
props.	—	properties
syst.	—	system
transluc.	—	translucent
transpar.	—	transparent

Minerals

Identification Tables and Plates

No.	Colour / Streak	Lustre / Transparency	Hardness / Specific Gravity	Cleavage Fracture & o. Phys. Props.	Common Form, Aggregates / Crystalline Syst.	Occurrence / Assoc. Minerals / Similar Minerals	Name & Chem. Formula / Origin of Specimen
1	silver-white / —	metallic / opaque	liquid / 13.6	— / greasy	little drops / trigonal	in ore veins, in cinnabar deps. / cinnabarite / —	nat. **MERCURY** (Quicksilver) Hg / Idrija, Slovenia Yugoslavia
2	tin-white, grey silver-white / grey	metallic / opaque	1.5—2 / 8.0—8.3	very good / brittle	tabular, lamellar, skeletal / monoclinic	in gold ore / sphalerite, pyrite, telluride / krennerite, nat. tellurium, calaverite	**SYLVANITE** AgAuTe₄ / Baia de Arieș, Romania
3	silver-white, greyish grey	metallic / opaque	2—3 / 7.6—7.8	— / sectile, malleable	granular masses, crystals / orthorhombic, pseudotetragonal	in ore veins / selenium ores / clausthalite	**EUCAIRITE** Cu₂Se.Ag₂Se / Săcărâmb, Romania
4	silver-white, reddish / light grey, silver-white	metallic / opaque	2—2.5 / 9.7—9.8	very good / brittle, malleable	granular, compact, lamellar, dendritic / trigonal	in ore veins / — / niccolite, cobaltite, breithauptite	nat. **BISMUTH** Bi / Krupka, Czechoslovakia
5 6	silver-white, grey to opaque / silver-white & shining	metallic / opaque	2.5—3 / 10.1 to 11.1	— / hackly, malleable, ductile	compact, thread-like, platy, curling, mossy / cubic	in ore veins / argentite, pyrargyrite, proustite, galena, pyrite, barytes, siderite, fluorite, calcite, quartz / dyscrasite, nat. antimony	nat. **SILVER** Ag / Kongsberg on Lågenelf, Norway / nat. **SILVER** Ag / Příbram, Czechoslovakia
7	silver-white, yellowish / yellowish grey	metallic / opaque	2.5—3 / 8.6	very good / fragile	compact, granular, lamellar / orthorhombic	in gold veins / — / sylvanite, calaverite, nat. tellurium	**KRENNERITE** (Au,Ag)Te₂ / Săcărâmb, Romania
8	tin-white to grey / grey	metallic / opaque	3—3.5 / 6.6	good / —	finely granular to coarse-spathic masses, compact, reniform / trigonal	in arsenic-silver ore veins / arsenic, antimonite, sphalerite, siderite, calcite / —	**ALLEMON-TITE** As+Sb / Příbram, Czechoslovakia

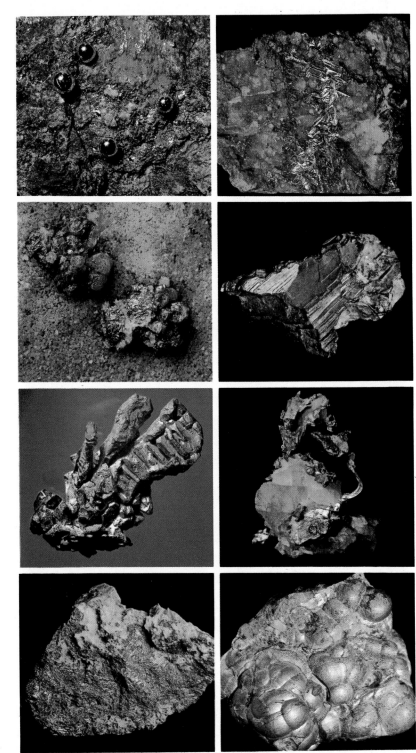

No.	Colour / Streak	Lustre / Transparency	Hardness / Specific Gravity	Cleavage / Fracture & o. / Phys. Props.	Common Form, Aggregates / Crystalline Syst.	Occurrence / Assoc. Minerals / Similar Minerals	Name & Chem. Formula / Origin of Specimen
9	silver-white to steel-grey / silver-white	metallic / opaque	4 – 4.5 / 14 – 19	— / hackly, ductile, malleable	grains, lumps, lamellae / cubic	in placer deps. / chromite, magnetite, ilmenite / nat. iron	nat. **PLATINUM** Pt / Nizhni Tagil, Central Urals, USSR
10	tin-white / grey-black	metallic / opaque	5 – 5.5 / 7.4	clear, good / brittle	compact, granular, nodular, fibrous, short-columnar, needle-shaped / orthorhombic	in ore veins / arsenopyrite / arsenopyrite	**LÖLLINGITE** $FeAs_2$ / Złoty Stok, Poland
11	tin-white / greyish-black	metallic / opaque	5 – 5.5 / 7.1 – 7.4	good / brittle	needle-shaped, fibrous crystals, compact nodules / orthorhombic	in ore veins / arsenopyrite / arsenopyrite	Variety of löllingite **LEUCO-PYRITE** (löllingite with reduced content of As) / Złoty Stok, Poland
12	silver-white, steel-grey / greyish-black	metallic / opaque	5 / 6.7	good / brittle, uneven	compact, granular / cubic	in ore veins / gersdorffite, calcite, siderite / gersdorffite, smaltite, galena	**ULLMANNITE** NiSbS / Monte Narba, Sardinia
13	silver-white, steel-grey / greyish-black	metallic / opaque	5 – 5.5 / 5.6 – 6.2	very good / brittle	granular, compact, spathic / cubic	in ore veins / ullmannite, chalcopyrite, siderite, calcite / ullmannite	**GERSDORF-FITE** NiAsS (Grey Nickel Pyrites, Nickel Glance) / Harzgerode, Harz, Germany
14	tin-white to light steel-grey / greyish black	metallic / opaque	5.5 – 6 / 6.4 – 6.9	— / brittle, uneven; jointing due to zonality	compact, reniform, massive, finely granular / cubic	in silver-cobalt-nickel ore veins / silver, safflorite, niccolite / safflorite, gersdorffite, cf. No. 390	**CHLOAN-THITE** $NiAs_3$ (White Nickel) / Schneeberg, Ore Mountains, Germany
15	tin-white to grey / black	metallic / opaque	5.5 – 6 / 5.9 – 6.2	clear, good / brittle	short-columnar, tabular crystals, compact, granular, radial, fibrous / monoclinic, pseudoorthorhombic	in ore veins / galena, silver, löllingite, chloanthite, skutterudite, cf. No. 389	**ARSENO-PYRITE** FeAsS (Arsenical Pyrite, Mispickel) / Freiberg, Ore Mountains, Germany
16	tin-white to grey / greyish-black	metallic / opaque	5.5 – 6 / 6.8	clear / brittle, uneven	cubic octahedral crystals, often bent, compact, reniform, massive, finely granular / cubic	in silver-cobalt-nickel ore veins / silver, safflorite, niccolite & o. / safflorite, löllingite, arsenopyrite, cf. No. 391	**SKUTTE-RUDITE** CoAs (Smaltite) / Bou Azzer, Morocco

9	10
11	12
13	14
15	16

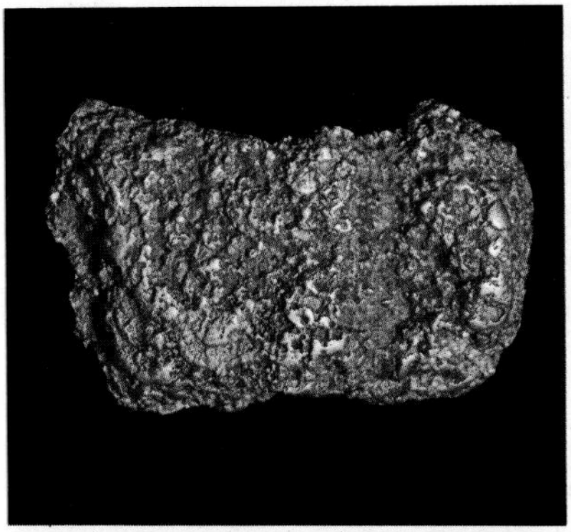

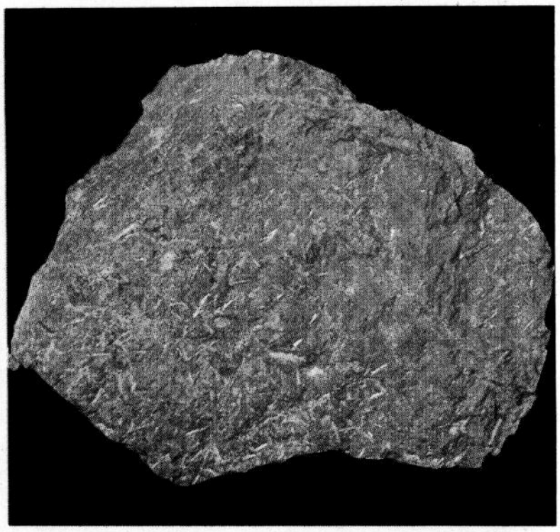

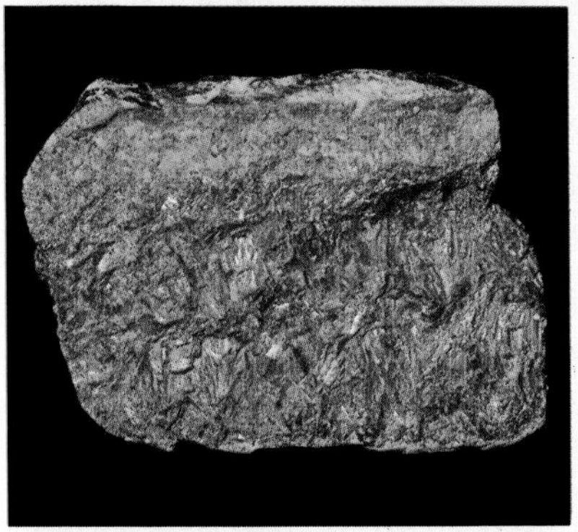

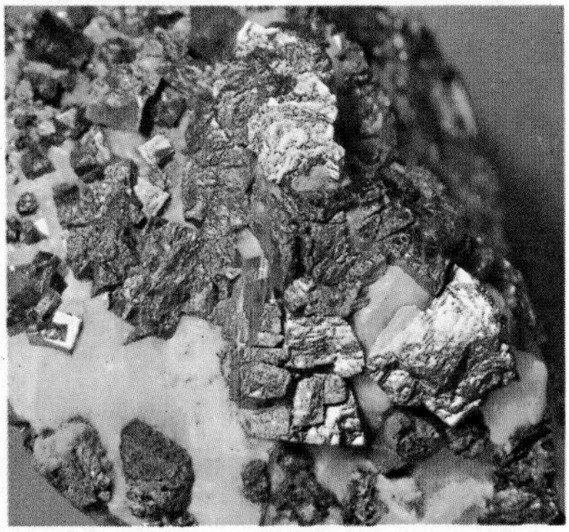

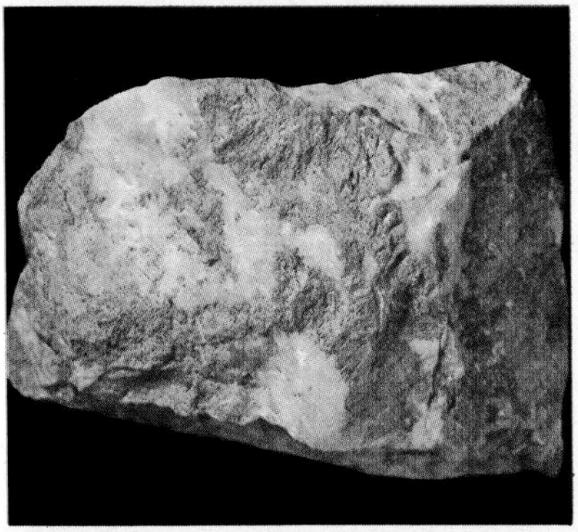

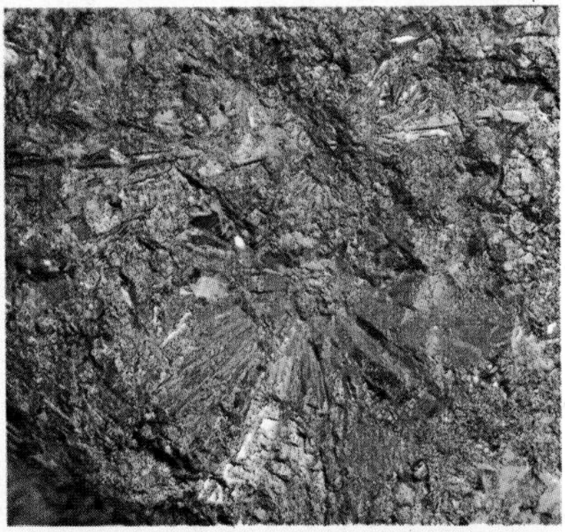

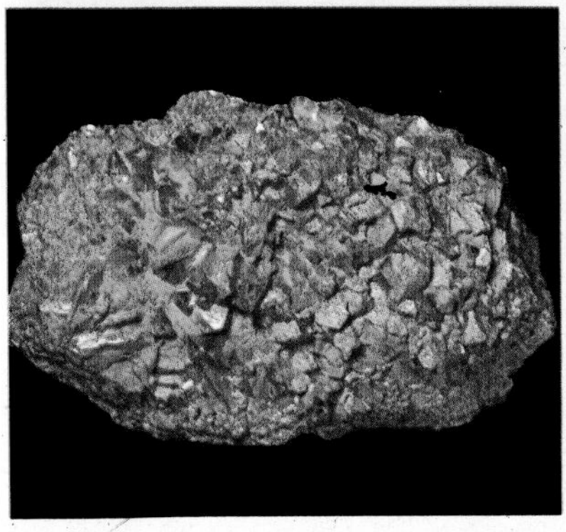

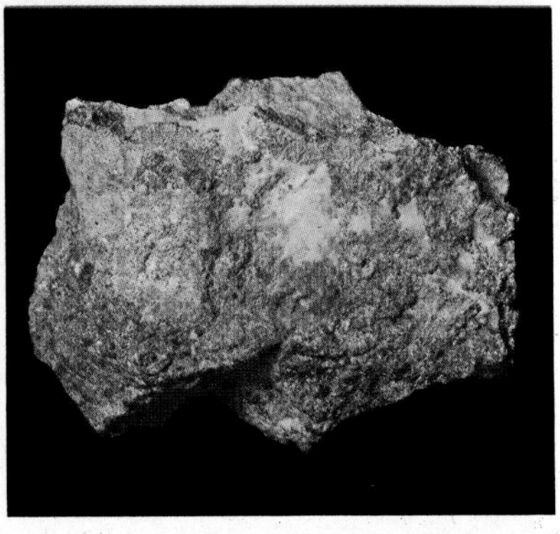

No.	Colour / Streak	Lustre / Transparency	Hardness / Specific Gravity	Cleavage / Fracture & o. / Phys. Props.	Common Form, Aggregates / Crystalline Syst.	Occurrence / Assoc. Minerals / Similar Minerals	Name & Chem. Formula / Origin of Specimen
17	colourless, white-yellowish / white	greasy / transpar., transluc.	1.0 — 2 / 1.5	poor / con-choidal, soft, water-soluble	skeletal crystals, crusty, botryoidal, earthy, fibrous, compact / cubic	in burning coal piles, in volcanoes / — / —	**AMMONIUM CHLORIDE** NH_4Cl / Vesuvius, Italy
18	white, yellowish-white / white	dull / transluc.	1 — 2 / 1.7	— / powdery, earthy	nodular, compact, earthy, scaly / monoclinic	in clays, marls, sandstones, gypsum, chalk / — / alunite	**ALUMINITE** $Al_2[(OH)_4SO_4] \cdot 7H_2O$ / Chuchle, Czechoslovakia
19	white, grey-white, yellowish, brownish, greenish, pink / white	dull / macro-scopically opaque	1 — 2 / 2 — 3	— / soft, friable, swelling in water	compact, massive, earthy, friable masses / monoclinic	in clays, in volcanic tuff / — / —	**MONTMORIL-LONITE** (Al,Mg) $[(OH)_2Si_4O_{10}] \cdot (Na,Ca)_x \cdot 4H_2O$ / Kuzmice, Czechoslovakia
20	white, grey, green, yellowish, brownish / white	pearly, greasy, dull / transluc., opaque	1 — 1.5 / 2.6 — 2.8	perfect / very soft, inelastic, flexible, greasy	lamellar crystals, scaly, massive, compact / monoclinic	alteration product of serpentine, in crystalline schists / chlorite, serpentine, magnetite, pyrite, dolomite, etc. / pyrophyllite, kaolinite	**TALC** Mg_3 $[(OH)_2Si_4O_{10}]$ (Soap-Stone, Steatite) / Providence, Rhode Island, USA
21	white, yellowish, greenish, bluish, reddish / white	pearly, dull / opaque	1 — 2 / 2.5 — 2.6	very good / soft, earthy	thin plates, fine-scaled, massive, loose, earthy / monoclinic	weathering product of felspars, in clays, as late formation in ore veins / quartz, mica / montmorillonite, pyrophyllite	**KAOLINITE** Al_4 $[(OH)_8Si_4O_{10}]$ / Podbořany, Czechoslovakia
22	white, yellowish grey-green, reddish / white	greasy / opaque	1 — 2 / 2.2 — 2.3	— / greasy	scaly, loose, earthy, massive, compact / monoclinic	as late formation in diff. rocks & serpentine / — / talc, pyrophyllite	**SAPONITE** $Mg_3[(OH)_2$ $AlSi_3O_{10}] \cdot nH_2O$ (Soap-Stone) / Faroes, Nólsoy, Denmark
23	white, grey, greenish, yellowish / white	pearly, dull / transluc.	1 — 2 / 2.6 — 2.9	perfect / soft, greasy	tabular crystals, rose-shaped, lamel-lar, scaly, massive / monoclinic	in quartz veins & ore veins, in slate clays / — / talc, kaolinite	**PYRO-PHYLLITE** Al_2 $[(OH)_2Si_4O_{10}]$ / Badin, North Carolina, USA
24	white, white-grey, yellowish, green-blue / white	dull / opaque	1.5 — 2 / 2.3	— / soft, floats on water	nodular masses, compact, massive, porous, earthy / orthorhombic	weathering product of serpentine / opal, chalcedony, magnesite, chlorite / —	**SEPIOLITE** Mg_4 $[(OH)_2Si_6O_{15}]$ $2H_2O + 4H_2O$ (Meerschaum) / Eskischehir, Turkey

41

No.	Colour — Streak	Lustre — Transparency	Hardness — Specific Gravity	Cleavage Fracture & o. Phys. Props.	Common Form, Aggregates Crystalline Syst.	Occurrence Assoc. Minerals Similar Minerals	Name & Chem. Formula Origin of Specimen
25	colourless, white, yellowish, brownish — white	vitreous to dull — transpar., transluc.	1.5 – 2 — 2.2 – 2.3	very good soft, brittle, water-soluble	granular, compact skeletal crystals — trigonal	isolated deps. — halite, mirabilite, epsomite, gypsum, clay — —	**NITRO-NATRITE** $NaNO_3$ (Chile Saltpetre, Soda Nitre) — Antofagasta, Chile
26	white — white	dull — opaque	1 – 2 — 2.6	— earthy, porous	porous, earthy masses, loose — trigonal	organic sediments — clay, flint — kaolinite, montmorillonite, cf. No. 51 to 54, 170, 319	Calcite variety **CHALK** $CaCO_3$ — Heligoland
27	white, yellowish — white	silky, vitreous, pearly — transluc.	1.5 — 1.7	very good soft, water-soluble	capillary, acicular crystals, reniform, fibrous, scaly, granular agg., crusts — triclinic	in ore veins, in coal piles, in clays — pyrite, melanterite alunite	**ALUNOGEN** $Al_2[SO_4]_3 . 18H_2O$ (Keramohalite) — Bolzano, Italy
28	white, yellowish, pink, greenish — white	silky — transluc.	1.5 — 1.6 – 1.7	good water-soluble	fibrous, acicular, capillary crystals, & agg. — monoclinic	in rocks as weathering product of sulphides — — alunogen	**APJOHNITE** $MnAl_2[SO_4]_4 . 22H_2O$ (Manganese Alum) — Alum Cave, Tennessee, USA
29	colourless, whitish — white	vitreous — transpar., transluc.	1 – 2 — 1.6	— brittle, conchoidal, water-soluble	fibrous, tubular agg., crusts; lamellar — cubic	in coal piles, in gypsum, clay, volcanoes — — —	**TSCHERMI-GITE** $NH_4Al[SO_4]_2 . 12H_2O$ (Ammonium Alum) — Čermíky near Kadaň, Czechoslovakia
30	colourless, white — white	vitreous to dull — transpar., transluc.	1.5 – 2 — 1.4 – 1.5	good conchoidal, water-soluble	tabular crystals, crusts, fibrous, earthy, granular, compact masses — monoclinic	as efflorescent crusts, in salt deps. — gypsum, halite, clay, marl — glauberite, glaserite	**MIRABILITE** $Na_2[SO_4] . 10H_2O$ (Glauber Salt) — Hallstatt, Salzkammergut, Austria
31	colourless, white, yellow, grey, red, brown — white	vitreous, silky, pearly — transpar., transluc.	1.5 – 2 — 2.3 – 2.4	very good soft, inelastic, flexible	columnar, tabular, needle-shaped, often twinned crystals, loose crystals, in druses, coarse to fine-grained, fibrous, scaly, massive, compact, earthy — monoclinic	rocks in salt deps., weathering product of sulphides in sedimentary rocks, in ore deps. — anhydrite, aragonite, sulphur	**GYPSUM** $CaSO_4 . 2H_2O$ — Duchcov, Czechoslovakia
32						anhydrite, mica, talc, kaolinite, cf. No. 132	**GYPSUM** $CaSO_4 . 2H_2O$ — Prešov, Czechoslovakia

43

No.	Colour / Streak	Lustre / Transparency	Hardness / Specific Gravity	Cleavage / Fracture & o. / Phys. Props.	Common Form, Aggregates / Crystalline Syst.	Occurrence / Assoc. Minerals / Similar Minerals	Name & Chem. Formula / Origin of Specimen
33 / 34	colourless, white, yellow, blue, violet, reddish, orange / white	vitreous transpar., transluc.	2 / 2.1 — 2.2	perfect, cube / soft to brittle, water-soluble	cubic crystals, coarse to fine-grained, compact, fibrous, massive / cubic	isolated deps., crusts in deserts, in volcanoes / anhydrite, gypsum, sylvite, dolomite & o. / sylvite, anhydrite, cf. No. 224	ROCK SALT NaCl (Halite) / Wieliczka near Cracow, Poland. / ROCK SALT NaCl (Halite) / Dognecea, Romania
35	colourless, white, yellowish, grey / white	vitreous, greasy / transpar., transluc.	2 / 1.9 — 2	very good, cube / soft to brittle, water-soluble	cubic, octahedral crystals, coarse to fine-grained, massive / cubic	sedimentary deps., as crusts in volcanoes / halite, anhydrite, carnallite, kainite & o. / rock salt	SYLVITE KCl / Stassfurt, Germany
36	colourless, white, greenish, yellowish, grey, red, brown / white	vitreous, greasy / transpar., transluc.	2 / 1.6	— / conchoidal, brittle, fusible, water-soluble	thick-tabular crystals, coarsely-granular masses, fibrous, compact / orthorhombic	in rock salt / anhydrite, kainite, kieserite & o. / kainite	CARNALLITE $KMgCl_3 \cdot 6H_2O$ / Stassfurt, Germany
37	colourless, white / white	vitreous to dull / transluc.	2 / 1.8	clear / brittle, uneven, weathered	tabular crystals, with orthorhombic habit / monoclinic	in boron deps. / colemanite & o. boron minerals / —	INYOITE $Ca[B_3O_3(OH)_5] \cdot 4H_2O$ / Mount Blanco on Furnace Creek, California, USA
38	colourless, white, yellowish / white	dull / transluc.	2 / 3.7	clear / —	very small octahedral crystals, crusts, needles, powdery, fibrous / cubic	as alteration product of arsenic minerals / — / —	ARSENOLITE As_2O_3 (White Arsenic) / Jáchymov, Czechoslovakia
39	colourless, grey, yellowish, brown, reddish, black / white, grey	adamantine, greasy / transpar., transluc.	1.5 — 3 / 5.5 — 5.6	— / sectile, flexible, ductile	cubic, octahedral crystals, compact, horny crusts, as impregnations / cubic	in silver veins / silver, cerussite, calcite, barytes, limonite / bromargyrite	KERARGYRITE AgCl (Chlorargyrite, Horn Silver) / Zacatecas, Mexico
40	colourless, white, reddish, brown / white	adamantine, pearly, greasy / transpar., transluc.	2.5 — 3 / 2.9	good / brittle, fissile	pseudocubic, short-columnar crystals, compact, granular, spathic, massive / monoclinic	in pegmatites / quartz, siderite, galena, pyrite & o. / anhydrite, barytes	CRYOLITE $Na_3[AlF_6]$ / Ivigtut, Greenland

No.	Colour / Streak	Lustre / Transparency	Hardness / Specific Gravity	Cleavage / Fracture & o. / Phys. Props.	Common Form, Aggregates / Crystalline Syst.	Occurrence / Assoc. Minerals / Similar Minerals	Name & Chem. Formula / Origin of Specimen
41	white, grey, yellowish / white	silky, dull / transluc.	2—2.5 / 3.2—3.8	very good / slightly brittle	cryptocrystalline, compact, reniform, massive, conchoidal, earthy, crusts / monoclinic	weathering product in zinc deps. / smithsonite, sphalerite / smithsonite	HYDRO-ZINCITE $Zn[(OH)_3CO_3]_2$ (Zinc Bloom) / San Giovanni Mine, Laorca, Sardinia
42	colourless, white, yellowish, grey / white	vitreous, dull / transpar., transluc.	2—3 / 1.9	very good / conchoidal, brittle, water-soluble	columnar, tabular crystals, rough planes, uneven / monoclinic	eliminated from soda lakes, in clays / — / —	GAYLUSSITE $Na_2Ca(CO_3)_2 . 5H_2O$ (Natrocalcite) / Sangerhausen, Germany
43	colourless, white, grey, yellowish / white	vitreous, greasy, dull / transpar., transluc.	2—2.5 / 1.7—1.8	clear soft, conchoidal, water-soluble	columnar, flat-tabular crystals, with dull crust; compact, earthy, powdery / monoclinic	on edges of borax lakes / halite, soda, mirabilite / kernite	BORAX $Na_2[B_4O_5(OH)_4]$ $8H_2O$ (Tincal) / Tibet
44	colourless, white, grey, yellowish, red / white	vitreous, greasy / transpar., transluc.	2.5—3 / 2.7—2.8	very good / brittle, conchoidal	tabular, columnar crystals, in druses, coarse spathic, compact, reniform, earthy / monoclinic	in saline deps. / natural potassium salts, anhydrite, gypsum / mirabilite	GLAUBERITE $CaNa_2[SO_4]_2$ / Villa Rubia de los Ojos, Spain
45	colourless, white, yellowish, reddish / white	vitreous, silky / transpar., transluc.	2—2.5 / 1.6	very good / non-hygro-scopic, water-soluble	needle-shaped, crystals, fibrous agg., earthy crusts / orthorhombic	as weathering product in ore deps., efflorescent crusts in steppes, alteration product of kieserite / — / kieserite	EPSOMITE $Mg[SO_4] . 7H_2O$ (Bitter Salt) / Kremnice, Czechoslovakia
46	colourless, white, yellowish, red, violet / white	vitreous / transpar., transluc.	2.5—3 / 2.1	perfect infusible, water-soluble	thick-tabular pseudorhombic crystals, mostly finely granular, compact masses / monoclinic	in salt deps. / halite, carnallite, kieserite, anhydrite / carnallite	KAINITE $KMg[ClSO_4] . 3H_2O$ / Stassfurt, Germany
47	white, grey-red, reddish-brown, yellowish / white	vitreous, pearly / transpar., transluc.	2.5—3 / 3	very good / uneven	small tabular & columnar crystals, compact, botryoidal masses / orthorhombic	in zinc ore deps. / hemimorphite, vanadinite / —	HOPEITE $Zn_3(PO_4)_2 . 4H_2O$ / Altenberg near Aachen, Germany
48	colourless, white, yellowish, reddish / white	vitreous, silky, dull / transluc.	2—2.5 / 2.6	very good / soft	fine needle-shaped, capillary crystals, powdery en-crustations; crusts; mammillary, botryoidal, reniform, earthy, powdery / monoclinic	efflorescent crusts on arsenic ores / niccolite, chloan-thite, erythrite, annabergite / —	PHARMA-COLITE $CaH[AsO_4] . 4H_2O$ / Jáchymov, Czechoslovakia

No.	Colour / Streak	Lustre / Transparency	Hardness / Specific Gravity	Cleavage / Fracture & o. / Phys. Props.	Common Form, Aggregates / Crystalline Syst.	Occurrence / Assoc. Minerals / Similar Minerals	Name & Chem. Formula / Origin of Specimen
49	white, grey, pink, brownish / white	adamantine, on cleavage planes pearly transluc.	2.5 / 5.6 — 5.8	very good brittle	needle-shaped, tabular, scaly crystals, compact, fibrous, radial, granular / orthorhombic	weathering product of antimony ores / antimonite, galena & o. / cerussite	VALENTINITE Sb_2O_3 (Antimony Bloom) / Cetinje, Yugoslavia
50	colourless / white	vitreous, pearly transpar.	2.5 / 2.2	good conchoidal	columnar, tabular crystals, often twinned in heart shape / monoclinic	in black coal seams, rare in ore deps. / — / calcite	WHEWELLITE $Ca[C_2O_4] . H_2O$ / Kladno, Czechoslovakia

Hardness 3 — 3.5

No.	Colour / Streak	Lustre / Transparency	Hardness / Specific Gravity	Cleavage / Fracture & o. / Phys. Props.	Common Form, Aggregates / Crystalline Syst.	Occurrence / Assoc. Minerals / Similar Minerals	Name & Chem. Formula / Origin of Specimen
51							CALCITE $CaCO_3$ (Calcareous Spar) / Helgustadir, Reydarfjördur, Iceland
52	colourless, white, yellow, red, brownish / white	vitreous, pearly transpar., transluc.	3 / 2.6 — 2.7	very good, rhombohedral brittle	columnar, tabular, needle-shaped, rhombohedral, scalenohedral crystals, rich in planes; in druses; coarse to fine-grained, spathic, stalactitic, reniform, compact, massive / trigonal	very common as crystals, in rock-fissures, in ore veins, deposited at hot springs, as sinter formation, as sedimentary rocks / — / dolomite, magnesite, ankerite, aragonite, chabasite, barytes, cf. No. 26, 170, 319	CALCITE $CaCO_3$ (Calcareous Spar) / Cumberland, England
53							CALCITE $CaCO_3$ (Calcareous Spar) / Egremont, Cumberland, England
54							CALCITE $CaCO_3$ (Calcareous Spar) / St. Andreasberg, Germany
55	colourless, white, yellowish / white	vitreous, pearly transpar., transluc.	3 / 2.7 — 2.8	very good brittle	rhombohedral crystals, finely granular masses, compact / trigonal	mixed crystals of calcite & cerussite, in lead ore deps. / — / calcite	PLUMBOCALCITE $(Pb,Ca)CO_3$ / Bleiberg, Carinthia, Austria
56	white, yellowish, grey / white	vitreous to dull transpar., transluc.	3 — 3.5 / 4.2	good brittle	columnar, tabular crystals, compact, botryoidal, granular, radial, lamellar, globular agg. & masses / orthorhombic, pseudohexagonal	in lead ore veins / galena, alstonite, barytes / cerussite, quartz	WITHERITE $BaCO_3$ / Alston Moor, Westmorland, England

No.	Colour / Streak	Lustre / Transparency	Hardness / Specific Gravity	Cleavage Fracture & o. Phys. Props.	Common Form, Aggregates / Crystalline Syst.	Occurrence / Assoc. Minerals / Similar Minerals	Name & Chem. / Formula / Origin of Specimen
57	colourless, white, grey, red, yellow, blue, brown / white	vitreous, pearly, greasy / transpar., transluc.	$\frac{3-3.5}{4.4}$	very good / brittle, conchoidal	thin- to thick-tabular, needle-shaped, pyramidal; lamellar, wisp-like, flabelliform, radial, reniform, granular agg.; massive, reniform, shelled / orthorhombic	frequent in ore veins, in fissures in sedimentary rocks, nodules in clays & marl / — / aragonite, calcite, anhydrite, felspars, cf. No. 144, 145, 234	**BARYTES** $Ba(SO_4)$ (Heavy Spar) / Baia Sprie (Felsöbánya), Romania
58	colourless, white, grey, bluish, violet, reddish / white	vitreous, pearly / transpar., transluc.	$\frac{3-3.5}{2.8-3}$	very good / distinct cleavage cracks, brittle	tabular, columnar, cube-shaped crystals, spathic, fibrous, compact, botryoidal, granular, massive / orthorhombic	in salt deps., from gypsum due to loss of water, forms 'Snake Alabaster' in ore veins, exhalation product in lava / halite, gypsum / cryolite, gypsum barytes, calcite, cf. No. 190	**ANHYDRITE** $Ca(SO_4)$ / Bochnia near Cracow, Poland
59	colourless, white, grey, yellow, reddish, brown / white	vitreous, pearly / transpar., transluc.	$\frac{3.5-4}{3}$	very good, rhombohedron / brittle, easily friable	rhombohedral, saddle-shaped crystals of twinned surface; coarse to fine-grained, spathic, massive, compact / trigonal	frequent in ore veins, as sedimentary rocks, metasomatic deps. in limestones / — / calcite, magnesite, cf. No. 151	**DOLOMITE** $CaMg(CO_3)_2$ / Banská Štiavnic Czechoslovakia
60	white, grey, yellowish, brownish / white	vitreous, pearly / —	$\frac{3.5-4}{3-3.1}$	very good, rhombohedron / brittle	rhombohedral, often twinned crystals, as in dolomite, compact, granular / trigonal	in siderite deps., in ore deps. / siderite, galena, sphalerite, quartz / siderite, breun-nerite	**ANKERITE** $CaFe[CO_3]_2$ / Mlýnky, Czechoslovakia
61	colourless, white, yellow, grey, reddish, violet, blue, green, brown / white	vitreous, greasy / transpar., transluc.	$\frac{3.5-4}{2.9-3}$	indistinct / conchoidal, brittle	prismatic, columnar crystals, often twinned, tubular-radiating, fragile agg. / orthorhombic	in rock-fissures, in ore deps., embedded in sulphur, as sinter formation / calcite, barytes, coelestine, strontianite, natrolite, topaz	**ARAGONITE** $CaCO_3$ / Hořenec, Czechoslovakia
62	white, yellowish / white	greasy / opaque	$\frac{3.5-4}{2.9-3}$	— / exfoliation (scaling)	pea-like / orthorhombic	deposited at hot springs / — / —	**ARAGONITE** $CaCo_3$ (Pisolite, Pea Stone, Pea Grit) / Carlsbad, Czechoslovakia
63	white / white	dull / transluc.	$\frac{3.5-4}{2.9-3}$	— / brittle	dendritic, ramifying agg. / orthorhombic	in siderite deps. / limonite / cf. No. 320, 485	**ARAGONITE** $CaCO_3$ (Flos Ferri) / Eisenerz, Styria, Austria
64	white, grey, yellow, green, pink / white	vitreous, greasy / transpar., transluc.	$\frac{3.5-4}{3.7}$	clear / brittle	pointed, fragile, needle-shaped crystals, fibrous, compact, radial / orthorhombic	in ore veins, in marl / calcite / aragonite, calcite, barytes, coelestine, natrolite	**STRONTIA-NITE** $SrCO_3$ / Sendenhorst, Germany

No.	Colour / Streak	Lustre / Transparency	Hardness / Specific Gravity	Cleavage Fracture & o. Phys. Props.	Common Form, Aggregates / Crystalline Syst.	Occurrence / Assoc. Minerals / Similar Minerals	Name & Chem. Formula / Origin of Specimen
65	white / white	dull / opaque	3.5 / 2.2	— / greasy	mammillary, nodular, radiating-columnar agg.; crusty, powdery / monoclinic	as weathering product of serpentine, embedded in volcanic tuffs / — / —	HYDRO-MAGNESITE $Mg_5[OH(CO_3)_2]_2 . 4H_2O$ / Vesuvius, Italy
66	colourless, white, grey / white	vitreous transpar., transluc.	3.5 — 4 / 2.4	very good / —	short- to long-columnar, tabular crystals, compact, granular, lamellar, massive / monoclinic	in clay & gypsum as nodules, lenses, in borax deps. / — / ulexite, pandermite, datolite	COLEMANITE $Ca[B_3O_4(OH)_3] . H_2O$ / Death Valley, California, USA
67	colourless, white, grey, yellowish, reddish / white	vitreous transluc.	3.5 — 4 / 2.7 — 2.8	good / brittle	rhombohedral, tabular crystals; tubular, fibrous, granular, compact, nodular, earthy / trigonal	isolated rocks, nodules in sedimentary rocks, in decomposed rocks / sulphide ores / aluminite	ALUNITE $KAl_3[(OH)_6(SO_4)_2]$ (Alum Stone) / La Tolfa near Civitavecchia, Italy
68	white, yellow, green, brown, bluish / white	vitreous, silky transluc.	3.5 — 4 / 2.3 — 2.4	good / brittle	fine-fibrous, radial bunches; semi-globular, mammillary, reniform agg. / orthorhombic	in fissures in sandstone, in silica schists, greywacke, quartzite / limonite, haematite / natrolite, prehnite, cf. No. 272	WAVELLITE $Al_3[(OH)_3(PO_4)_2] . 5H_2O$ / Cerhovice, Czechoslovakia
69	colourless, white, yellowish, bluish, brown / white	pearly, greasy to dull / opaque	3.5 — 4 / 1.9	exfoliation (scaling) / very brittle, conchoidal	shelly, globular, reniform, botryoidal structure / amorphous	in cavities in limonite deps. / — / chalcedony, some opals	EVANSITE $Al_3[(OH)_6PO_4] . 6H_2O$ / Železník, Czechoslovakia
70	colourless, white, yellowish, grey, reddish / white	vitreous, pearly, dull / transluc.	3.5 — 4 / 2.2 — 2.3	very good / less brittle, in the air dull & crumbly	long-columnar, needle-shaped crystals; tubular, compact, fibrous, earthy / monoclinic	in ore veins, in cavities & fissures in eruptive rocks / other zeolites, calcite, chlorite / felspars	LAUMONTITE $Ca[AlSi_2O_6]_2 . 4H_2O$ (Fibrous Zeolite) / Peter's Point, New Scotland, Canada
71	colourless, white, grey, red, brown / white	vitreous, pearly transpar., transluc.	3.5 — 4 / 2.2	very good / brittle	tabular, lamellar, short-columnar crystals, scaly, spathic, compact, flabelliform / monoclinic	in cavities & fissures in volcanic rocks, in ore deps., in crystalline schists / other zeolites / desmine, felspar, cf. No. 194	HEULANDITE $Ca[Al_2Si_7O_{18}] . 6H_2O$ (Stilbite, Lamellar Zeolite) / Teigarhorn, Berufjördur, Iceland
72	white, reddish, yellowish, brownish / white	vitreous, pearly, silky transpar., transluc.	3.5 — 4 / 2.1 — 2.2	very good / brittle	tabular, columnar, chip-shaped crystals; bundle-shaped bunches; radial agg. / monoclinic	in vesicles & druses in eruptive rocks, druses in granite, in alpine fissures, in ore deps. / other zeolites, calcite, quartz / other zeolites	DESMINE $Ca[Al_2Si_7O_{18}] . 7H_2O$ (Stilbite, Bundle Zeolite) / Teigarhorn, Berufjördur, Iceland

COLOURLESS AND WHITE MINERALS OF NON-METALLIC LUSTRE

No.	Colour Streak	Lustre Transparency	Hardness Specific Gravity	Cleavage Fracture & o. Phys. Props.	Common Form, Aggregates Crystalline Syst.	Occurrence Assoc. Minerals Similar Minerals	Name & Chem. Formula Origin of Specimen
73	colourless, yellow, green, violet, blue, pink / white	vitreous transpar., transluc.	4 / $3-3.2$	perfect, octa-hedron conchoidal, brittle, shining when heated	cubic, octahedral crystals, coarsely to finely granular, spathic, massive, compact, radial, striated / cubic	frequent in rock fissures & ore veins, isolated masses, metasomatic in limestone, as cementing material in sandstones / — / barytes, apatite, amethyst, cf. No. 212, 233	FLUORITE CaF_2 (Fluor Spar, Blue John) / Pare in Val Sarentina, Italy
74	white, grey-yellow / white	vitreous, greasy transluc.	$4-4.5$ / $3.6-3.7$	good / —	pointed, pyramidal, short-columnar crystals; fibrous, granular, massive / orthorhombic	in lead ore deps. / witherite, baryto-calcite, galena / —	ALSTONITE $CaCO_3 + BaCO_3$ (Bromlite) / Alston Moor, England
75	colourless, white, reddish, yellowish / white	vitreous, transpar., transluc.	$4-4.5$ / $2.1-2.2$	good brittle, striated cleavage planes	short-needle-shaped, thick-tabular crystals; pseudorhombic to pseudocubic poly-synthetic twins / monoclinic	in cavities in volcanic rocks, in deep-sea deps. / analcime, chabazite / harmotome	PHILLIPSITE $KCa[Al_3Si_5O_{16}] \cdot 6H_2O$ (Cross Stone) / Vinařická hora, Czechoslovakia
76	colourless, white, grey, reddish, yellowish / white	vitreous transpar., transluc.	4.5 / $2.4-2.5$	good brittle, smooth cleavage planes	thick-tabular, thick-columnar crystals; penetration twins as in phillipsite / monoclinic	in cavities in alkaline effusive rocks, in ore veins / barytes, calcite, galena, quartz / phillipsite	HARMO-TOME $Ba[Al_2Si_6O_{16}] \cdot 6H_2O$ (Cross Stone) / Strontian, Argyll, England
77	colourless, white, yellow, reddish / white	vitreous, transpar., transluc.	$2-2.1$	indistinct brittle	cube-shaped, rhombohedral crystals; penetra-tion twins, only in druses; no agg. / trigonal	in cavities in igneous rocks; in alpine clefts & ore veins / other zeolites, quartz, felspar / calcite, dolomite	CHABAZITE (Ca,Na_2) $[Al_2Si_4O_{12}] \cdot 6H_2O$ (Cubic Zeolite) / Řepčice, Czechoslovakia
78	colourless, white, yellowish, reddish, greenish / white	vitreous transpar., transluc.	4 / 2.1	poor, indistinct brittle, con-choidal	thick-tabular crystals; penetration twins / orthorhombic	in cavities in basaltic rocks / — / chabazite	LEVYNITE $Ca[Al_2Si_4O_{12}] \cdot 6H_2O$ (Zeolite) / Benyevenagh, Ireland
79	colourless, white, bluish, greenish, yellowish, grey, brown / white	vitreous, pearly, silky transpar., transluc.	$4.5-5$ / $3.3-3.5$	very good con-choidal, fibrous, brittle	tabular to needle-shaped, chip-shaped crystals; mammillary, reniform agg., crusts, massive / orthorhombic	in limestone & dolomite / galena, sphalerite, smith-sonite, wulfenite, calcite, limonite / prehnite, smith-sonite, chalcedony, opal, phosphorite	HEMI-MORPHITE Zn_4 $[(OH)_2Si_2O_7] \cdot H_2O$ (Silicate of Zinc, Galmei, Calamine) / Nertschinsk, Transbaikal, USSR
80	colourless, white, yellowish, grey / white	vitreous, silky transluc.	$4.5-5$ / 2.8	very good brittle	thick-tabular, needle-shaped, capillary, spathic, tubular to fibrous, lamellar / monoclinic	in contact-meta-morphic limestones in crystalline schists / quartz, garnet, vesuvianite, pyroxene / pektolite, tremolite	WOLLA-STONITE $Ca_3[Si_3O_9]$ (Table Spar) / Mirsk, Poland

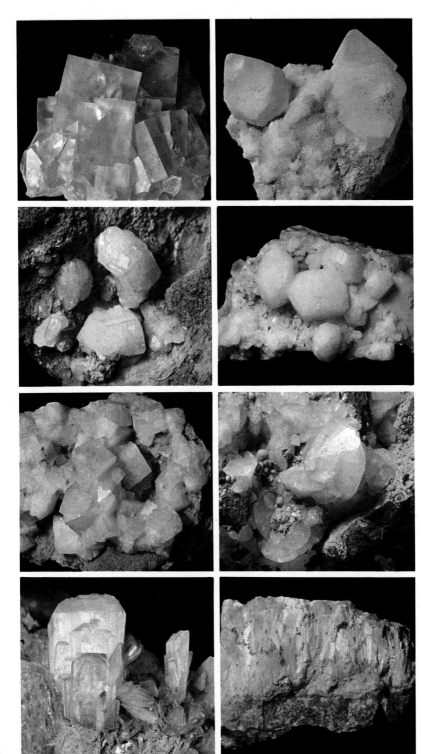

COLOURLESS AND WHITE MINERALS
OF NON-METALLIC LUSTRE

Hardness 5 –

No.	Colour / Streak	Lustre / Transparency	Hardness / Specific Gravity	Cleavage / Fracture & o. / Phys. Props.	Common Form, Aggregates / Crystalline Syst.	Occurrence / Assoc. Minerals / Similar Minerals	Name & Chem. Formula / Origin of Specimen
81	colourless, white, yellowish, reddish, brown / white	vitreous, greasy / transluc.	5 / 2.9–3	indistinct / brittle	thick-tabular, columnar crystals; radial agg. / orthorhombic	in weathering zone of lead deps. / – / aragonite	TARNOWITZITE (Pb,Ca)CO$_3$ / Tsumeb, South-West Africa
82	white, whitish-grey / white	greasy, dull / opaque	5 / 3.1	indistinct / conchoidal, brittle	columnar, hexagonal crystals; coarsely granular masses / hexagonal	large crystals in apatite veins in phosphate-pegmatites / – / cf. No. 158, 214, 281	APATITE here HYDROXYL-APATITE Ca$_5$[OH(PO$_4$)$_3$] / Snarum, Norway
83	white, greyish / white	vitreous, silky, pearly / transluc.	5 / 2.8	very good / very brittle	needle-shaped, long-columnar crystals; fibrous agg., compact, massive, reniform, radiating / triclinic	in fissures in igneous rocks / zeolite, calcite / tremolite, wollastonite	PECTOLITE Ca$_2$NaH[Si$_3$O$_9$] / Bergen Hill (Tunnel), New Jersey, USA
84	colourless, white, greyish, yellowish / white	vitreous, silky / transluc.	5 / 2.2–2.4	good / brittle	acicular, capillary crystals; radial, fine-fibrous, semi-globular agg., compact, massive, earthy / monoclinic	in fissures & cavities, in basic igneous rocks / other zeolites, calcite / natrolite, thomsonite	MESOLITE Na$_2$Ca$_2$ [Al$_2$Si$_3$O$_{10}$] . 8H$_2$O (Zeolite) / Hauenštejn, Czechoslovakia
85	colourless, white, yellow, grey, reddish / white	greasy, vitreous / transpar., transluc.	5–5.5 / 3–3.1	– / conchoidal	short-columnar crystals with longitudinal striation, mostly compact / monoclinic	hydrothermal formation in quartz veins, in vesuvian lavas / quartz, magnesite, siderite, barytes, apatite / –	WAGNERITE Mg$_2$[F PO$_4$] / Werfen, Salzburg, Austria
86	colourless, white, greyish, yellowish / white	vitreous, greasy / transpar., transluc.	5–5.5 / 2.9–3	indistinct / conchoidal	thick-tabular, short-columnar crystals, compact, granular, reniform, massive, crusty / monoclinic	in druses & fissures in basic igneous rocks, in granitic veins, in ore deps. / – / danburite, colemanite, zeolite	DATOLITE CaB[OH SiO$_4$] / Seiser Alm, South Tyrol Austria
87	colourless, white, dim, green-grey, grey, blue / white	vitreous, pearly / transpar., transluc.	5–6 / 2.7	good to clear / conchoidal brittle	short-prismatic, thick-columnar, tubular crystals; granular, fibrous, radial, massive / tetragonal	in metamorphic rocks, as volcanic ejecta / garnet, epidote, vesuvianite, pyroxene / felspars, spodumene, cf. No. 421	Scapolite variety MEIONITE Ca$_8$ [(Cl$_2$,SO$_4$,CO$_3$)$_2$ (Al$_2$Si$_2$O$_8$)$_6$] / Vesuvius, Italy
88	colourless, white, yellowish, reddish, pink, red / white	vitreous, silky / transpar., transluc.	5–5.5 / 2.2–2.4	very good / brittle	columnar, acicular, capillary crystals; radial, globular, reniform, concentric, shelled, fibrous, compact, massive, powdery / orthorhombic	in cavities in igneous rocks, in fissures in granites & crystalline schists / other zeolites, calcite / aragonite, scolezite, thomsonite, mesolite, wavellite, cf. No. 199	NATROLITE Na$_2$Al$_2$Si$_3$O$_{10}$. 2H$_2$O (Fibrous Zeolite) / Česká Lípa, Czechoslovakia

No. 81–88

Table 1

No.	Colour / Streak	Lustre / Transparency	Hardness / Specific Gravity	Cleavage / Fracture & o. / Phys. Props.	Common Form, Aggregates / Crystalline Syst.	Occurrence / Assoc. Minerals / Similar Minerals	Name & Chem. Formula / Origin of Specimen
89	colourless, white, yellowish, reddish, grey / white	vitreous, dull / transpar., transluc.	5—5.5 / 2.1—2.3	— / conchoidal, uneven	isolated grown up crystals; druses; also compact, granular, massive, earthy / cubic	in vesicles & fissures in basic igneous rocks, in ore veins / zeolite, calcite / leucite, sodalite, garnet	**ANALCITE** $Na[AlSi_2O_6]$. H_2O / Seiser Alm, South Tyrol, Austria
90	white, yellowish, reddish, greenish, brown / white	vitreous / transpar., transluc.	5—5.5 / 2.2—2.4	good / brittle	acicular, short-columnar, thick-tabular crystals; reniform, globular, nodular agg.; radial, tubular bunches; compact / orthorhombic	in vesicles in basic igneous rocks, in vesuvian lavas / other zeolites, analcite, calcite / natrolite, prehnite	**THOMSONITE** $NaCa_2$ $[Al_2(Al,Si)$ $Si_2O_{10}]_2$. $6H_2O$ (Comptonite, Fibrous Zeolite) / Doubice, Czechoslovakia
91	colourless, white, yellowish, brownish / white	vitreous, silky / transpar., transluc.	5—5.5 / 2.1—2.4	very good / brittle, conchoidal	columnar, needle-shaped to capillary crystals; bunchy, radial, fibrous agg.; also compact / monoclinic	in cavities & fissures in volcanic rocks, in cracks in granite & syenite, in crystalline schists, in alpine cracks / other zeolites, calcite / thomsonite, aragonite	**SCOLEZITE** $Ca[Al_2Si_3O_{10}]$. $3H_2O$ (Fibrous Zeolite) / Teigarhorn, Berufjördur, Iceland
92	colourless, white, grey / white	vitreous, greasy to dull / transluc., mostly dull, opaque	5.5—6 / 2.5	— / conchoidal	isometric, rounded, embedded crystals; seldom grown up, granular agg. / tetragonal to pseudocubic	constituent of volcanic rocks, in vesuvian lavas & ashes / — / analcite, sodalite, garnet	**LEUCITE** $K[AlSi_2O_6]$ / Ariccia, Italy
93	colourless, white, blue, grey, greenish / white	vitreous, greasy / transpar., transluc.	5.5—6 / 2.3—2.5	good / conchoidal	seldom isometric crystals; mostly compact, granular agg. / cubic	constituent of diff. igneous rocks, in volcanic ejecta / — / leucite, analcite, hauyne, nosean, lazurite, cf. No. 237	**SODALITE** Na_8 $[Cl_2(AlSiO_4)_6]$ / Vesuvius, Italy
94	colourless, white, grey, yellow, dark-green / white	vitreous / transpar., transluc.	6 / 3.7—3.9	very good / —	octahedral, cubic, grown up & embedded crystals; compact, granular / cubic	in metamorphosed limestones & dolomites, in volcanic ejecta / — / —	**PERICLASE** MgO / Monte Somma, Vesuvius, Italy
95	colourless, water-clear, white / white	vitreous / transpar., transluc.	5.5—6 / 2.1	— / conchoidal	botryoidal, reniform, conical, platy, massive, crusty / amorphous	in cracks & cavities in volcanic rocks, in ore veins, deposited at hot springs / — / —	**HYALITE** SiO_2 . nH_2O (Glassy Opal) / Valeč, Czechoslovakia
96	white, enamel-like / white	dull / opaque	5.5—6 / 1.9—2	— / conchoidal	reniform, botryoidal, massive, porcellanic / amorphous	in vesicles & cavities in basalt rocks / — / —	Opal variety **KASCHO-LONG** SiO_2 . nH_2O / Hüttenberg, Carinthia, Austria

No.	Colour Streak	Lustre Transparency	Hardness Specific Gravity	Cleavage Fracture & o. Phys. Props.	Common Form, Aggregates Crystalline Syst.	Occurrence Assoc. Minerals Similar Minerals	Name & Chem. Formula Origin of Specimen
97	milk-white, yellowish-white _white_	dull opaque	5 — 6 (also softer) 2.1 — 2.3	— con-choidal	veins, nodules, opaline amorphous	base opal containing alum, in hornstones or jasper — —	ALUMO-CALCITE $SiO_2 . nH_2O + Al_2O_3$ Eibenstock, Ore Mountains, Germany
98	colourless, greenish, yellowish _white_	vitreous transpar.	6 2.6	— con-choidal, brittle	columnar, hexagonal, grown up crystals hexagonal	in pegmatites, in fissures in granites smoky quartz, adularia, apatite, titanite aquamarine	MILARITE $KCa_2 [(Be,Al)Si_4O_{10}]_3$ Val Giuv, Graubünden, Switzerland
99	colourless, white _white_	vitreous, pearly transpar., transluc.	6 2.5	very good —	tabular crystals, frequent twin striation monoclinic	in fissures in syenite zeolite zeolite, felspar	EUDI-DYMITE NaBe $[OH Si_3O_7]$ Langesunds-fjord, Norway
100	white, reddish, yellowish _white_	vitreous, pearly transluc.	6 2.6	very good brittle, uneven	thick-tabular, columnar, acicular crystals, often with twin striation; compact, granular, radial, spathic monoclinic	as rock constituents in granites, syenites, pegmatites, in crystalline schists, in fissures in gneiss, in ore veins — barytes	ALBITE $Na[AlSi_3O_8]$ (Felspar Family) Intschitobel near Amsteg, Switzerland
101	colourless, white, greenish _white_	vitreous, pearly transpar., transluc.	6 2.5	very good brittle	only grown up thick-tabular crystals, often covered with green chlorite dust monoclinic	in druses in alpine cracks, in ore veins quartz, albite, pericline, titanite, chlorite —	ADULARIA $K[AlSi_3O_8]$ (Felspar Family) Grimsel, Switzerland
102	colourless, white, grey-white, reddish _white_	vitreous, pearly transpar., transluc., dim	6 2.7	very good brittle	tabular, short-columnar, embedded & grown up crystals; massive, granular triclinic	rock constituents in metamorphic limestone & dolomite, in ore deps., in volcanic ejecta — cf. No. 204	ANORTHITE $Ca[Al_2Si_2O_8]$ (Felspar Family) Monte Somma, Vesuvius, Italy
103	colourless, white, greenish, reddish, yellowish _white_	vitreous, pearly transpar., transluc., opaque	6 2.6 — 2.7	very good brittle	tabular crystals; massive, granular agg.; compact, granular, massive triclinic	rock constituent in diorites, andesites, syenites, in cordie-rite-gneiss — —	ANDESINE Constituent of $Na[AlSi_3O_8]$ $Ca[Al_2Si_2O_8]$ (albite + anorthite) (Felspar Family) Esterel, France
104	colourless, white, green _white_	vitreous, pearly transpar., transluc.	6 2.6	very good brittle	grown up twinned plagioclases; columnar, thick-tabular crystals; no agg. triclinic	in druses in alpine cracks quartz, adularia, titanite, rutile, anatase, chlorite —	PERICLINE $Na[AlSi_3O_8] + K, Ca$ (Felspar Family) Rothenkopf, Zillertal, Tyrol, Austria

No.	Colour Streak	Lustre Transparency	Hardness Specific Gravity	Cleavage Fracture & o. Phys. Props.	Common Form, Aggregates Crystalline Syst.	Occurrence Assoc. Minerals Similar Minerals	Name & Chem. Formula Origin of Specimen
105	colourless, whitish, bluish —— white	pearly, vitreous with bright bluish shine transluc.	6 —— 2.5 – 2.6	very good brittle	thick-tabular, wide-columnar crystals; spathic —— monoclinic	due to differentiation of pale milky adularia or sanidine —— – —— –	MOONSTONE K[(Si,Al)$_4$O$_8$] (Felspar Family) —— Kandy, Ceylon
106	white, yellow, brown —— whitish	vitreous transluc.	6 – 6.5 —— 3.1 – 3.2	indistinct —	small, short-columnar crystals, rich in planes; compact, granular agg. —— orthorhombic	in contact metamorphic deps., in volcanic ejecta, in ore deps. —— – —— cancrinite	HUMITE Mg$_7$[(OH,F)$_2$ (SiO$_4$)$_3$] —— Monte Somma, Vesuvius, Italy
107	colourless, white, yellowish, reddish, violet, grey —— white	vitreous, pearly —— transpar., transluc.	6.5 – 7 —— 3.3 – 3.5	very good, lamellar conchoidal, very brittle	tabular, wide-columnar, foliaceous, acicular crystals; compact, lamellar, radial, tubular, scaly —— orthorhombic	constituent in bauxite, in crystalline schists —— kyanite, corundum, magnetite, calcite, serpentine —— gypsum	DIASPORE α-AlOOH —— Banská Štiavnica, Czechoslovakia
108	colourless, yellowish-white, greenish, grey —— white	vitreous transpar., transluc.	6.5 – 7 —— 3.1 – 3.2	very good brittle	wide-tubular, columnar, thick-tabular, spathic —— monoclinic	in pegmatites —— lepidolite, quartz, albite, tourmaline, beryl —— felspars, scapolite	SPODUMENE LiAl[Si$_2$O$_6$] —— Chesterfield, South Carolina, USA
109	colourless —— white	vitreous, greasy transpar.	7 —— 2.6	— conchoidal	columnar to needle-shaped grown up crystals in druses —— trigonal	in druses & fissures in magmatic rocks & crystalline schists, in ore veins —— albite, titanite, rutile & o. —— phenakite, topaz, beryl	QUARTZ SiO$_2$ (Rock Crystal) —— Bourg d'Oisans, Dauphiné, France
110	milk-white, also differently coloured —— white	vitreous, greasy transpar., transluc.	7 —— 2.6	— conchoidal	columnar, embedded & grown up crystals; mostly horizontally striated, coarsely to finely granular —— trigonal	very frequent constituent of diff. rocks, in ore veins —— – —— –	QUARTZ SiO$_2$ (Milky Quartz) —— Ratibořice, Czechoslovakia
111	white, yellowish, reddish, grey, brownish —— white	vitreous transluc., opaque	7 —— 2.6	— conchoidal	radiating-columnar agg. —— trigonal	quartz variety with morphological peculiarities, in vesicles in basalts —— – —— –	ASTERIATED QUARTZ SiO$_2$ —— Peřimov, Czechoslovakia
112	white, grey, yellow, red, brown, greenish, blackish —— white	greasy, waxy —— transluc., opaque	5.5 – 6.5 —— 1.9 – 2.5	— conchoidal, brittle	veins, nodules, concretions, filling of cavities, disseminated, compact, botryoidal, nodular, reniform —— amorphous	product of decomposition of many rocks, in ore veins, deposited at hot springs —— – —— –	OPAL SiO$_2$. nH$_2$O —— Madeira

05	106
07	108
09	110
11	112

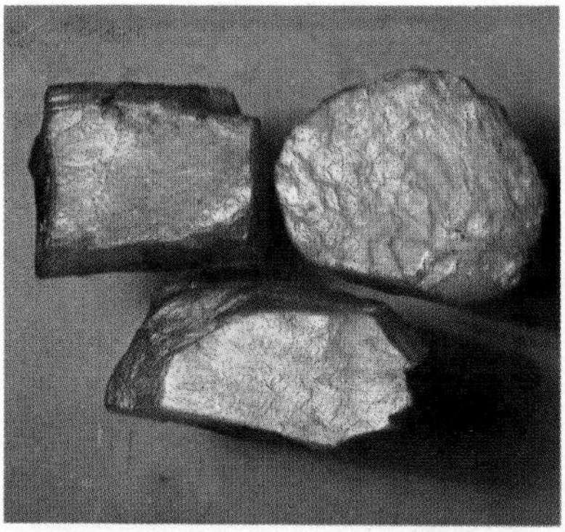
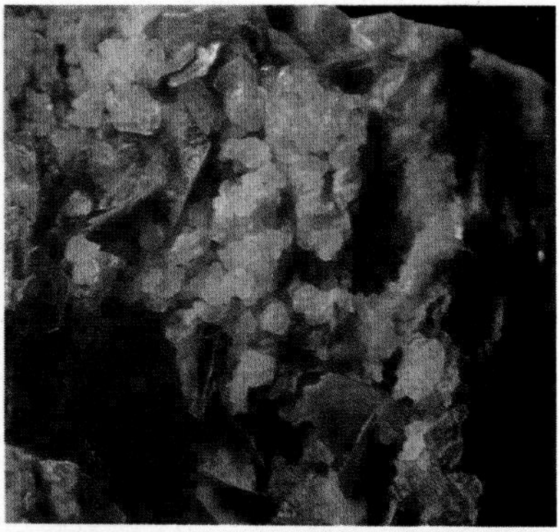
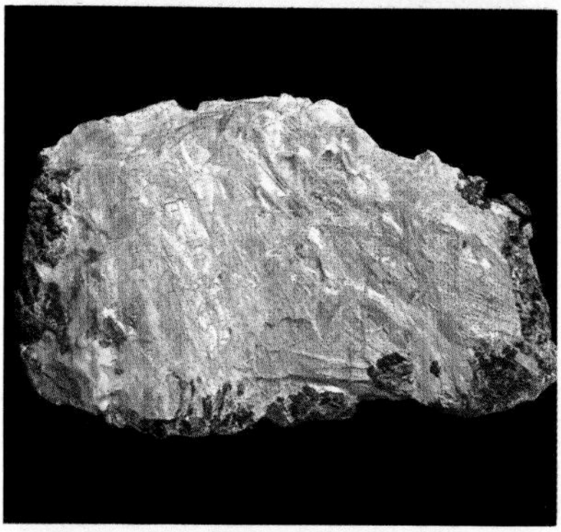
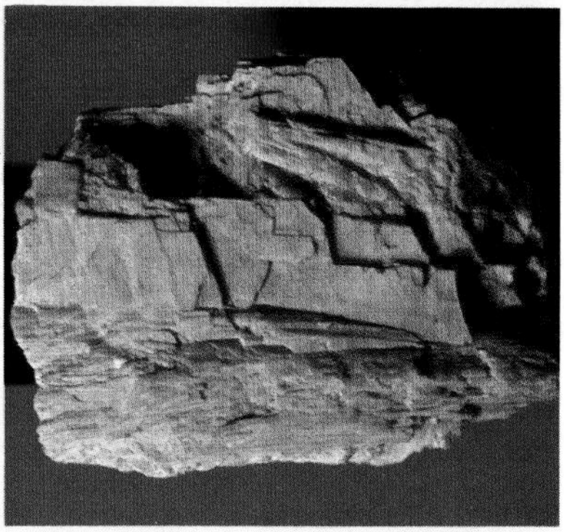
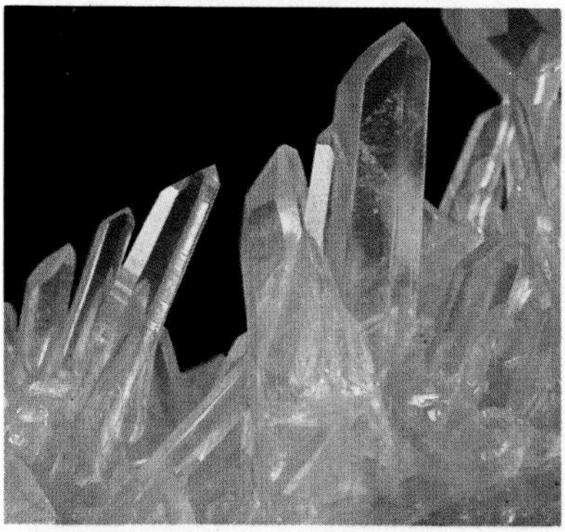
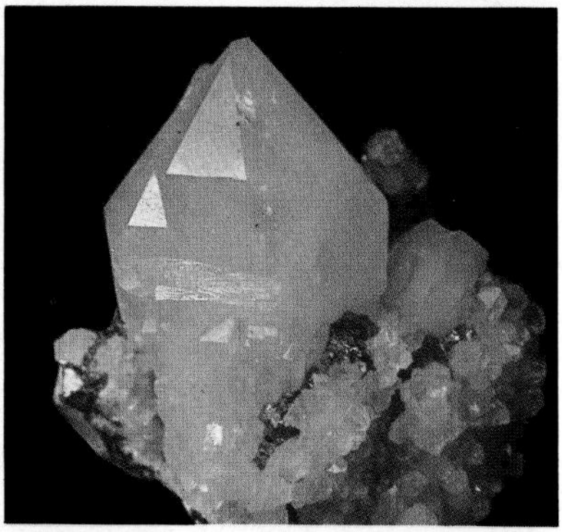
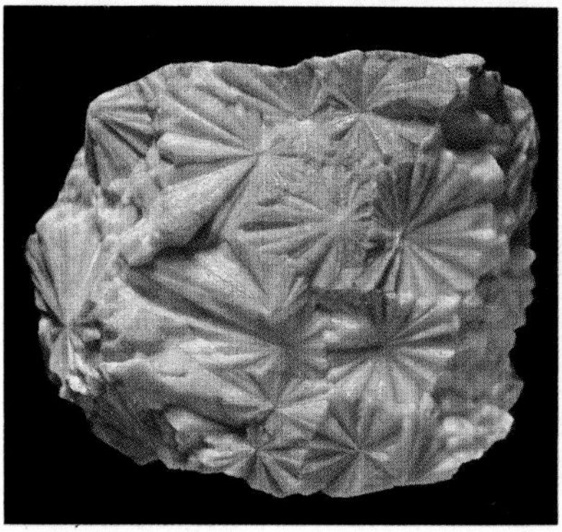
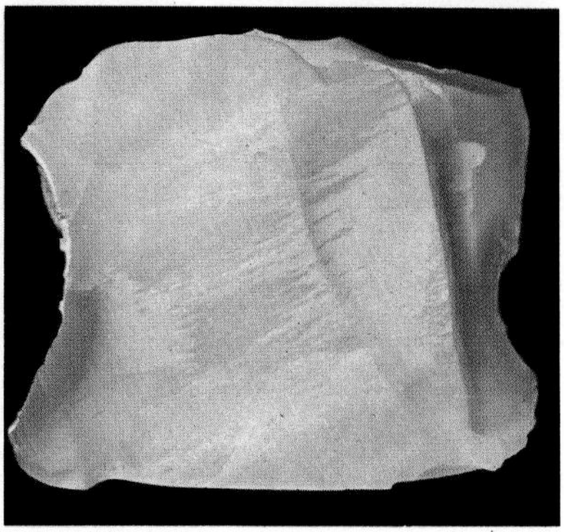

No.	Colour Streak	Lustre Transparency	Hardness Specific Gravity	Cleavage Fracture & o. Phys. Props.	Common Form, Aggregates Crystalline Syst.	Occurrence Assoc. Minerals Similar Minerals	Name & Chem. Formula Origin of Specimen
113	colourless, white, yellowish, grey	vitreous, pearly, also dull transpar., transluc.	7	indistinct	thin-tabular crystals, flabelli-form twins & triplets	in cavities in trachyte, andesite, porphyry	**TRIDYMITE** SiO_2
	white		2.2	brittle	orthorhombic	—	Colli Euganei, Italy
						—	
114	colourless, grey, yellowish, greenish, bluish	vitreous, greasy or ada-mantine transpar., transluc.	7	—	cube-shaped em-bedded crystals; compact, massive, fibrous nodules	in potash salt deps.	**BORACITE** $Mg_3[Cl\,B_7O_{13}]$ (Stassfurtite)
	white		2.9 — 3	con-choidal	orthorhombic, pseudotetragonal	embedded in gypsum, anhydrite, carnallite	Lüneburg, Germany
						—	
115	colourless, grey-white, yellowish	vitreous transpar., transluc.	7.5	very good	thick-tabular, vertically striated, many-planed crystals	in syenites & pegmatites in Nor-way & Madagascar	**HAMBER-GITE** $Be_2[(OH,F)\,BO_3]$
	white		2.3	brittle	orthorhombic	—	Imalo, Madagascar
116	colourless, white, yellow, pink, brown	vitreous transpar., transluc.	7.5 — 8	less clear to good	hexagonal, quartz-like, short-colum-nar, grown up crystals	in beryllium-peg-matites, in cracks in granite & mica slates, in tin ore deps.	**PHENAKITE** Be_2SiO_4
	white		3	brittle, con-choidal	trigonal	beryl, apatite, topaz, tourmaline, quartz	San Miguel di Piracicaba, Minas Gerais, Brazil
						quartz, topaz	
117	colourless, yellow, blue, green, red, reddish	vitreous transpar., transluc.	8	very good	short- to long-columnar crystals with orthorhombic outline, as impregna-tion; compact, mas-sive, felspar-like	pneumatolytic struc-ture, in pegmatites, granites, in tin ore veins, in placers of precious stones	**TOPAZ** $Al_2[F_2SiO_4]$
	white		3.5 — 3.6	uneven, con-choidal	orthorhombic	—	Tonokamiyama, Japan
						phenakite, quartz, corundum, beryl, aragonite, cf. No. 168, 523	
118	colourless, yellow, green, blue, red, black, grey	ada-mantine, greasy, also dull transpar., transluc., opaque	10	very good	octahedral, cubic, dodecahedral crystals; bulbous & striated planes frequent; granular, radiate, globular agg.	embedded in ultra-basic rocks, in con-glomerates, loose crystals in placers, in river sands	**DIAMOND** C
	white		3.5	con-choidal, brittle	cubic	—	Kimberley, South Africa
						cf. No. 513	

YELLOW MINERALS OF METALLIC LUSTRE *Hardness 2.5 — 3*

No.	Colour Streak	Lustre Transparency	Hardness Specific Gravity	Cleavage Fracture & o. Phys. Props.	Common Form, Aggregates Crystalline Syst.	Occurrence Assoc. Minerals Similar Minerals	Name & Chem. Formula Origin of Specimen
119	gold-yellow, light yellow, brownish, mustard-yellow	metallic opaque		—	compact, sheets, grains, dendritic skeleta; octahedral, cubic, dodecahedral crystals rare; feather-shaped, arborescent, fili-form, capillary, platy agg., mossy	in hydrothermal veins in volcanic rocks, lamellar, granular & nodular in placers	**nat. GOLD** Au
			2.5 — 3			quartz, pyrites, galena, arseno-pyrite, antimonite, gold-telluride	Roșia Montană, Romania
			14.5 to 19.3	ductile, malleable, flexible, hackly			**nat. GOLD** Au
120	yellow, golden, shining				cubic	pyrites, chalco-pyrite, marcasite	Jílové, Czechoslovakia

No.	Colour Streak	Lustre Transparency	Hardness Specific Gravity	Cleavage Fracture & o. Phys. Props.	Common Form, Aggregates Crystalline Syst.	Occurrence Assoc. Minerals Similar Minerals	Name & Chem. Formula Origin of Specimen
121	gold-yellow, light yellow, mustard-yellow, brownish	metallic opaque	2.5 – 3 14.5 to 19.3	— ductile, malleable, flexible, hackly	sheets, plates, grains, compact, dendritic skeleta; rarely octahedral, cube-like, dodeca-hedral crystals; feather-shaped, ar-borescent, filiform, capillary, platy agg., mossy cubic	in hydrothermal veins, in volcanic rocks, lamellar, granular, in lumps in placers quartz, pyrites, galena, arseno-pyrite, antimonite, gold-telluride pyrites, chalco-pyrite, marcasite	nat. GOLD Au Křepice, Czechoslovakia nat. GOLD Au Roşia Montană, Romania
122	yellow, golden, shining						
123	speiss-yellow, brass-yellow, brownish greyish-black	metallic opaque	3 – 3.5 5.5	very good brittle	acicular, capillary crystals in bunches; radiate, fibrous agg.; granular, compact trigonal	in ore veins, filling cracks in coal seams siderite, chalco-pyrite, barytes, fluorite, calcite —	MILLERITE NiS (Hair Pyrites) Kladno, Czechoslovakia
124	brass-yellow, often tarnished greenish-black to black	metallic opaque	3.5 – 4 4.1 – 4.3	indistinct soft, brittle, uneven	isometric crystals; octahedral & tetra-hedral; often striated; compact, reniform, finely granular, massive tetragonal, pseudocubic	very frequent in diff. ore deps. pyrites, galena, sphalerite, carbon-ate, quartz pyrites, gold, marcasite	CHALCO-PYRITE CuFeS$_2$ (Copper Pyrites) Freiberg, Ore Mountains, Germany

Hardness 4 – 6.5

No.	Colour Streak	Lustre Transparency	Hardness Specific Gravity	Cleavage Fracture & o. Phys. Props.	Common Form, Aggregates Crystalline Syst.	Occurrence Assoc. Minerals Similar Minerals	Name & Chem. Formula Origin of Specimen
125	gold-yellow, brass-yellow, occasion-ally tarnished greenish-black	metallic opaque	6 – 6.5 4.8 – 5	indistinct con-choidal, brittle	well-developed em-bedded & grown up octahedral, cube-like, pentagondo-decahedral crystals; cube planes often striated; compact, granular, reniform, nodular, closely packed fibres, radiate, massive cubic	very frequent in sulphidic ore deps., in sedimentary rocks, as mixture in magmatic rocks, crystalline schists, in coals, as petri-faction medium in sedimentary rocks — marcasite, chalco-pyrite, gold, pyrrhotite	PYRITES FeS$_2$ (Sulphuric Pyrites, Iron Pyrites) Rio Marina, Elba PYRITES FeS$_2$ (Sulphuric Pyrites, Iron Pyrites) Tavistock, Devon, England
126							
127	brass-yellow to greenish greenish-black	metallic opaque	6 – 6.5 4.8 – 4.9	indistinct brittle, in damp air weathers easily	tabular, short-columnar, acicular, chip-shaped crys-tals; pectiniform, fibrous agg.; mas-sive, compact, reni-form, nodular, radial, globular, stratified orthorhombic	with pyrites in ore deps., in lignites, marls, clays, not in igneous rocks, less frequent than pyrites — pyrites, chalco-pyrite, pyrrhotite, gold	MARCASITE FeS$_2$ (Spear Pyrites) Komořany, Czechoslovakia
128	yellow-brown, dark-brown, occasion-ally tarnished grey-black	metallic opaque	4 4.6	clear (fissile) brittle, uneven	thick-tabular to short-columnar crystals; rose-shaped agg., dis-seminated, com-pact, granular, massive hexagonal	in basic igneous rocks, in meta-morphic rocks, in crystalline schists, in pegmatites, in ore veins pyrites, magnetite, chalcopyrite & o. pyrites, chalco-pyrite, marcasite, cf. No. 308	PYRRHO-TITE (Pyrrhotine) FeS (Magnetic Pyrites) Herja (Chiuz-baia, Kisbánya), Romania

No.	Colour / Streak	Lustre / Transparency	Hardness / Specific Gravity	Cleavage / Fracture & o. / Phys. Props.	Common Form, Aggregates / Crystalline Syst.	Occurrence / Assoc. Minerals / Similar Minerals	Name & Chem. Formula / Origin of Specimen
129	yellow, orange-yellow, yellow-brown, brownish	adamantine, greasy transpar., transluc.	$1-2.5$ / 2	indistinct / very brittle, conchoidal, easily inflammable, burns with blue flame	pyramidal, tabular, many-planed crystals; compact, lamellar, massive, granular, reniform, earthy, powdery / orthorhombic	in sedimentary limestones & marls; as sublimate in volcanoes & hot springs, on coal piles / calcite, aragonite, gypsum, anhydrite, rock salt / —	nat. SULPHUR S / Cianciana, Sicily, Italy
130	straw-yellow						nat. SULPHUR S / Agrigento, Sicily, Italy
131	yellow, lemon-yellow / lemon-yellow	adamantine, greasy, vitreous, pearly transpar., transluc.	$1.5-2$ / $3.4-3.5$	very good / sectile, soft, flexible	lamellar, columnar, tabular, short-prismatic crystals; compact, platy, tubular, radial, reniform, nodular / monoclinic	in ore veins, disseminated in clays, marls, on volcanoes, at hot springs / realgar, antimonite, sphalerite, pyrites / —	ORPIMENT As_2S_3 (Tinsel) / Allchar near Mrežičko, Macedonia, Yugoslavia
132	yellow, brownish / white	silky, pearly transluc.	$1.5-2$ / $2.3-2.4$	very good / soft	coarse to fine-fibrous, parallel agg. / monoclinic	in cracks in sedimentary rocks, in salt deps. / — / cf. No. 31, 32	SELENITE $Ca[SO_4] . 2H_2O$ (Satin Spar) / Sverdlovsk, Central Urals, USSR
133	sulphur-yellow, lemon-yellow, orange-yellow / yellow	dull, silky transluc.	$1-3$ / 2.5	good / —	capillary, acicular, lamellar, tabular crystals; compact, fine-fibrous, scaly, powdery, earthy; flabelliform encrustations, efflorescences / monoclinic	decomposition product of uranium minerals besides other uranium micas / — / uranophane	ZIPPEITE $[6UO_2 3(OH) 3SO_4] . 12H_2O + 3H_2O$ (Uran Ochre) / Jáchymov, Czechoslovakia
134	yellowish-white, straw-yellow / yellowish, straw-yellow	dull, silky transluc.	2 / $4-4.5$	clear / —	efflorescences, encrustations, fibrous, earthy, radial, disseminated / orthorhombic	weathering product of molybdenite & other molybdenum minerals / — / different ochres	MOLYBDITE MoO_3 or $Fe_2(MoO_4)_3 . 7H_2O$ (Molybdenum Ochre) / Buena Vista, Colorado, USA
135	reddish, yellow, orange, yellowish-brown / straw-yellow	dull opaque	$2.5-3$ / $3.9-4.5$	— / conchoidal	compact, massive, reniform, disseminated, gummous, loose, earthy / amorphous	weathering product of uraninite / — / —	GUMMITE $(Pb,Ca,Ba)_2 [O_8 SiO_4] . 5H_2O$ (Uranium Ochre) / Flat Rock Mine, Carolina, USA
136	yellowish-green, dirty yellow, light yellowish / pale yellow	pearly transpar., transluc.	$2-3$ / 3.5	very good / —	foliated, needle-shaped, scaly crystals; lamellar, radial, bunchy agg.; crusty / tetragonal	in pegmatites, in quartz veins / uranium minerals / autunite, cf. No. 138, 260	URANO-CIRCITE $Ba[UO_2 PO_4]_2 . 8H_2O$ (Uranium Mica) / Falkenstein, Vogtland, Germany

No.	Colour / Streak	Lustre / Transparency	Hardness & o. / Specific Gravity	Cleavage / Fracture & o. / Phys. Props.	Common Form, Aggregates / Crystalline Syst.	Occurrence / Assoc. Minerals / Similar Minerals	Name & Chem. Formula / Origin of Specimen
137	light yellow, greenish-yellow / pale yellow, greenish-yellow	vitreous, pearly transluc.	2.5 / 2.5	very good / —	thin, tabular, micaceous crystals; fine-scaly agg.; earthy, encrustations / triclinic, pseudohexagonal	weathering product of uranium minerals / — / —	SCHROECK-INGERITE $NaCa_3[UO_2 F SO_4 (CO_3)_3] . 10H_2O$ (Uranium Mica) / Jáchymov, Czechoslovakia
138	sulphur-yellow, lemon-yellow, yellowish-green, greenish-yellow / yellowish, greenish	pearly, vitreous transluc.	2–2.5 / 3–3.2	very good / soft	thin- to thick-tabular, grown up crystals; scaly clusters, lamellar agg.; massive, earthy / tetragonal	as secondary uranium mineral in granites & pegmatites, in cracks, in uranium ore veins / quartz, fluorite / other uranium micas, cf. No. 260	AUTUNITE $Ca[UO_2 PO_4]_2 . 10H_2O$ (Lime Uranite) / Sabugal, Portugal
139	colourless, white, yellow, greenish, brown, apple-green / white	silky transluc.	2.5 / 1.8–2	— / water-soluble	capillary, fibrous, acicular crystals / monoclinic	weathering product of pyrites in ore deps., in lignites / — / apjohnite	HALO-TRICHITE $FeAl_2[SO_4]_4 . 22H_2O$ (Iron Alum, Feather-Alum, Hair Salt) / Dubník, Czechoslovakia
140	honey-yellow, orange, brownish, yellowish-white / whitish	greasy transpar., transluc., dull	2–2.5 / 1–1.1	— / conchoidal, brittle, electric when rubbed	gravel, pebbles, plates, drop-shaped, disseminated / amorphous	fossil resin in sedimentary rocks ('Bernsteinerden') along East- & North-Sea coast / — / cf. No. 552	AMBER $C_{12}H_{20}O$ (Succinite) / Klaipeda, USSR
141	straw-yellow, orange-yellow, brownish / whitish	greasy transluc., opaque	2–2.5 / 1	— / brittle, conchoidal	disseminated, massive, banded, nodular, stratified / amorphous	amber-like fossil resin from brown coal, lignite / — / —	WALCHO-VITE $(C_{15}H_{26}O)_4$ / Valchov, Czechoslovakia
142	colourless, yellow, red, brown, green, grey, black / white, yellow, brown	adamantine, greasy, almost metallic transpar. to opaque	3.5–4 / 3.9–4.2	very good, rhombo-dodecahedron / brittle	tetrahedral, dodecahedral, cubic, often twinned crystals; coarse to fine-grained, massive, compact, fibrous, shelled-reniform masses / cubic	in ore veins, hydrothermal, pneumatolytic, also sedimentary / galena, chalcopyrite, quartz & o. / tetrahedrite, cassiterite, garnet, cf. No. 191, 315, 445	SPHALERITE ZnS (Blende, Black Jack) / Banská Štiavnica, Czechoslovakia
143	yellow, orange, brown / yellow	adamantine, (only crystals), dull transluc.	3–3.5 / 4.9–5	good / brittle	rarely short-columnar crystals; mostly as powdery coating; massive, earthy / hexagonal	in weathering zone of zinc deps. / sphalerite, galena, smithsonite / zippeite, bismutite	GREEN-OCKITE CdS / Bishoptown near Paisley, Scotland
144	colourless, white, yellow, grey, red, blue, brown / white	vitreous, greasy, pearly transpar., transluc.	3–3.5 / 4.4	very good / less brittle, conchoidal	thin- to thick-tabular, columnar, needle-shaped, pyramidal, lamellar, radiated, reniform, granular agg.; / orthorhombic	in ore veins, in cracks in sedimentary rocks / — / aragonite, calcite, cf. No. 57, 145, 234	BARYTES $Ba[SO_4]$ (Heavy Spar) / Baia Sprie (Felsöbánya), Romania

No.	Colour Streak	Lustre Transparency	Hardness Specific Gravity	Cleavage Fracture & o. Phys. Props.	Common Form, Aggregates Crystalline Syst.	Occurrence Assoc. Minerals Similar Minerals	Name & Chem. Formula Origin of Specimen
145	colourless, white, yellow, red, grey, blue, brown white	vitreous, pearly, greasy transpar., transluc.	3—3.5 4.4	very good less brittle, conchoidal	tabular, columnar, lamellar crystals, also massive, granular orthorhombic	very common in ore veins — — cf. No. 57, 144, 234	BARYTES Ba[SO₄] (Heavy Spar) Příbram, Czechoslovakia
146	white, yellow, orange, brown, greenish, grey white	adamantine, greasy transluc.	3.5 7.1	good brittle, conchoidal	columnar, thick-tabular, barrel-shaped, needle-shaped; reniform agg.; earthy, crusty hexagonal	in lead & manganese ore deps. arsenic ores, limonite wad, psilomelane, quartz pyromorphite, cf. No. 174	MIMETESITE Pb₅[Cl(AsO₄)₃] (Mimetite) Johanngeorgenstadt, Ore Mountains, Germany
147	yellow, orange-yellow, brownish white	adamantine, greasy transluc.	3.5—4 7—7.2	— brittle	botryoidal, reniform coatings; compact, globular agg. hexagonal	arsenical pyromorphite of Nussière — —	Variety of pyromorphite NUSSIERITE Pb₅Cl(PO₄)₃ + As Beaujeu, Dep. Rhône, France
148	yellow, ochre-yellow, brown straw-yellow	silky, resinous transluc.	3—4 2.3—2.8	— brittle	acicular, capillary crystals; fibrous, radial agg.; reniform, globular, compact, earthy hexagonal	in iron ore veins haematite, siderite, limonite different ochres	KAKOXEN (Cacoxenite) Fe₄[(OH) PO₄]₃ . 12H₂O Hrbek, Czechoslovakia
149	yellow, green-yellow light yellow	dull transluc., opaque	4 4.4—4.5	very good —	tabular, lamellar crystals; flabelliform agg.; powdery, earthy, reniform, botryoidal coatings monoclinic	impregnation in sandstones & limestones, secondary uranium mineral — autunite	CARNOTITE (K,Na,Ca,Pb)₂ [(UO₂)₂ V₂O₈] . 3H₂O Sabugal, Portugal
150	yellow, greenish-yellow to brown light yellow	vitreous, pearly, dull transluc.	3—4 4.6	very good —	small pyramidal crystals; mostly compact, earthy, crusty orthorhombic	secondary uranium mineral — curite & torbernite —	SODDITE (UO₂)₁₅[(OH)₂₀ Si₆O₁₇] . 8H₂O Shinkolobwe, Katanga, Zaïre
151	white, yellow, reddish, brown white	vitreous, pearly transpar., transluc.	3.5—4 3	very good brittle, easily friable	rhombohedral crystals; coarsely to finely granular, spathic, massive trigonal	in ore veins, metasomatic from limestone — — cf. No. 59	DOLOMITE CaMg[CO₃]₂ Banská Štiavnica, Czechoslovakia
152	white, yellow, light brown, brown white	vitreous, pearly transluc.	4 3.4	good brittle	flat, lentiform, rhombohedral crystals; also massive, spathic, granular trigonal	mixed crystals between magnesite & siderite, in siderite ore deps. — siderite, magnesite, ankerite	MESITITE (Fe,Mg)CO₃ (Mesitine Spar) Traversella, Piedmont, Italy

No.	Colour Streak	Lustre Transparency	Hardness Specific Gravity	Cleavage Fracture & o. Phys. Props.	Common Form, Aggregates Crystalline Syst.	Occurrence Assoc. Minerals Similar Minerals	Name & Chem. Formula Origin of Specimen
153	white, yellow, grey, greenish — white	vitreous transpar., transluc.	4 / 3.7	very good brittle	small prismatic crystals in druses; granular, tubular agg.; spathic, massive — monoclinic	in ore veins — barytes, galena, fluorite, quartz	BARYTO-CALCITE BaCa[CO$_3$]$_2$ — Alston Moor, Westmorland, England
154	whitish, yellow, yellow-brown, brown-red — yellowish, brownish	dull opaque	5.5−6 / 4.1−5.8	—	earthy crusts, powdery, compact masses, coatings, encrustations — cubic	weathering product of antimonite & o. antimony ores — other ochres	ROMEITE (Ca,NaH) Sb$_2$O$_6$ (O,OH,F) (Antimony Ochre) — St. Marcel, Piedmont, Italy
155	honey-yellow, ochre-yellow — white	greasy, resinous transluc.. opaque	5.5−6 / 1.9−2.5	— conchoidal, brittle	solid masses, veins, nodular — amorphous	opal displaying annual ring marking	WOOD OPAL SiO$_2$. nH$_2$O — Clover Creek, Lincoln Co., Idaho, USA
156	yellow, red-orange, brown-yellow, orange-yellow	ada-mantine, dull transluc.	4−5 / 7.2	good —	columnar, needle-shaped crystals; mostly compact, massive, granular, acicular, compact — orthorhombic	as alteration product of uranium ores & o. uranium minerals — —	CURITE 3PbO . 8UO$_3$. 4H$_2$O — Shinkolobwe, Katanga, Zaïre
157	white, yellow, brown, greenish, reddish — white	ada-mantine, greasy transpar., transluc.	4.5−5 / 5.9−6.1	clear brittle, con-choidal	pseudooctahedral, pyramidal, tabular, grown up & embedded crystals; compact, granular, crusty, as impregnation — tetragonal	in pegmatites, in tin deps., in meta-somatic limestone & dolomite, in lead veins — wolframite, molybdenite, quartz & o. anglesite, cerussite	SCHEELITE Ca[WO]$_4$ (Calcium Tungstate) — Obří důl, Giant Mountains, Czechoslovakia
158	colourless, white, yellow, greenish, violet, red — white	vitreous, greasy transpar., transluc.	5 / 3.1	clear con-choidal, brittle	hexangular, tabular, short- to long-columnar, many-planed crystals; compact, reniform, botryoidal, nodular, globular, massive, radial, granular — hexagonal	as constituent in magmatic rocks, in cracks, in crystalline schists, in pegmatic-pneumatolytic veins, in sedimentary deps. — beryl, quartz, nepheline, felspars, cf. No. 82, 214, 281	APATITE Ca$_5$[F(PO$_4$)$_3$] — Durango, Mexico
159	whitish, yellowish, greenish — white	greasy transluc.	4−4.5 / 2.8−3.1	— con-choidal	radial, fibrous, nodular, globular, botryoidal agg.; coatings, crusty, earthy — hexagonal	in cracks in weathered basalt rocks — chalcedony	STAFFELITE Ca$_5$[F(PO$_4$, CO$_3$OH)$_3$] (Francolite) — Staffel/Lahn, Germany
160	white, yellowish, greenish, greyish — white	dull opaque	4.5 / 2.5	— brittle	semi-globular agg., reniform crusts — orthorhombic	in cracks & cavities of sedimentary rocks & sedimentary iron ores — wavellite	BARRAN-DITE (Al,Fe)PO$_4$. 2H$_2$O (Aluminiferous Strengite) — Zbirov, Czechoslovakia

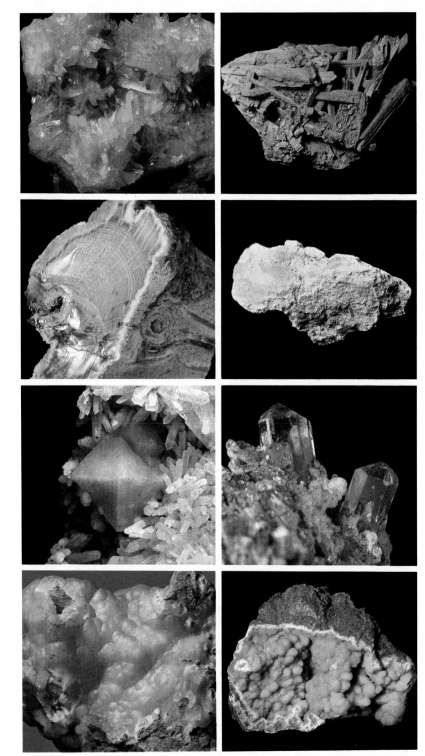

No.	Colour Streak	Lustre Trans- parency	Hard- ness Specific Gravity	Cleavage Fracture & o. Phys. Props.	Common Form, Aggregates Crystalline Syst.	Occurrence Assoc. Minerals Similar Minerals	Name & Chem. Formula Origin of Specimen
161	colourless, white, yellow, brownish white	vitreous transpar., transluc.	5.5—7 2.2	— conchoidal, brittle	subvitreous, amorphous, agglomerated grains of sand, compact amorphous	natural silica glass melted in volcanoes & thunderbolts (fulgurite), in meteorite craters — —	LECHATE- LIERITE SiO_2 Lybian Desert
162	white, yellow, brown, reddish greenish, black white	vitreous, greasy, adamantine transpar., transluc.	5—5.5 3.4—3.6	clear conchoidal, brittle	columnar, wedge-shaped, tabular, needle-shaped crystals; rarely compact, granular monoclinic	in alpine cracks, in crystalline schists, in magnesite deps., embedded in syenite & phonolite, as constituent of magmatic rocks — axinite, cf. No. 339, 463	TITANITE $CaTi[OSiO_4]$ (Sphene) Tavetsch, Bündner Oberland, Switzerland
163	straw-yellow, yellow-green, yellow-brown white	silky transluc.	5—5.5 2.9	good, clear brittle	fine-fibrous, capillary, fibrous bunches, entangled agg., compact, radiate orthorhombic	in druses in tin ore deps., in crystalline schists, in quartz — —	CARPHOLITE $MnAl_2$ $[(OH)_4 Si_2O_6]$ Slavkov, Czechoslovakia
164	yellowish, grey, whitish white	vitreous, pearly transluc., opaque	4.5—7 3.5—3.7	very good brittle	tubular, needle-shaped, lamellar agg. triclinic	cyanite in colourless to yellowish, radial entangled agg. — sillimanite	Cyanite variety RHAETIZITE $Al_2[OSiO_4]$ (Disthene) Alto Adige, Italy
165	yellow, white, grey, green white	vitreous, greasy transpar., transluc., dim	6—7 3.2—3.3	good, clear conchoidal	prismatic, thick-tabular, short-columnar crystals; massive, granular, nodular orthorhombic	constituent in basic rocks, in volcanic ejecta, in metamorphosed limestones —	FORSTERITE $Mg_2[SiO_4]$ Monte Somma, Vesuvius, Italy
166	honey-yellow, brown, green white	greasy, vitreous transluc., opaque, dim	6—6.5 3.1—3.2	poor, indistinct conchoidal	tetrahedral embedded & grown up crystals; globular agg. cubic	in ore deps., in syenite-pegmatites nepheline, sodalite, garnet, epidote vesuvianite, hessonite	HELVITE $(Mn,Fe,Zn)_8$ $[S_2 (BeSiO_4)_6]$ Schwarzenberg, Ore Mountains, Germany
							Hardness 7—8
167	yellowish, yellow white	vitreous, greasy transpar.	7 2.6	— conchoidal	columnar, grown up crystals & druses trigonal	colour variety of quartz — cf. No. 109	CITRINE SiO_2 Rio Grande do Sul, Brazil
168	straw-yellow, yellowish, greenish white	vitreous transluc.	8 3.5	very good conchoidal	parallel-radial, tubular agg. orthorhombic	tubular topaz in tin ore deps. wolframite, zinnwaldite, quartz, cassiterite & o. aragonite, cf. No. 117	Topaz variety PYCNITE $Al_2[F_2 SiO_4]$ Altenberg, Ore Mountains, Germany

No.	Colour / Streak	Lustre / Transparency	Hardness / Specific Gravity	Cleavage / Fracture & o. / Phys. Props.	Common Form, Aggregates / Crystalline Syst.	Occurrence / Assoc. Minerals / Similar Minerals	Name & Chem. Formula / Origin of Specimen
169	orange, light brown, brownish-black / yellow	greasy / opaque	2.5 / 1.8 — 2	— / conchoidal	globular, reniform nodules & masses / amorphous	in weathering zones of iron ore deps. / — / diadochite	DELVAUXITE $Fe_2[(OH)_3 PO_4]$. 3.5 H_2O / Nučice, Czechoslovakia
170	orange, brownish / white	dull / transluc.	3 / 2.6 — 2.7	good / brittle	botryoidal, stalactitic agg. / trigonal	in fissures & cavities in sedimentary limestones / — / —	CALCITE $CaCO_3$ (Calc-spar) / Mořina, Czechoslovakia
171	yellowish, orange, red, grey, brown, green / white	ada-mantine, greasy / transluc.	3 / 7.9 — 8.3	indistinct / brittle to soft, conchoidal	pyramidal, short-columnar crystals; bundle-shaped, globular / tetragonal	in tin ore deps. / cassiterite, scheelite, wolframite, quartz & o. / scheelite, cf. No. 323	STOLZITE $Pb[WO_4]$ / Cínovec, Czechoslovakia
172 / 173	orange, yellow, red, grey, greenish / white, light grey	ada-mantine, greasy / transpar., transluc.	3 / 6.5 — 6.9	clear / brittle to soft, conchoidal	tabular, short-columnar, cube-shaped, pyramidal, grown up crystals; compact, granular, crusty / tetragonal	in oxidation zone of lead ore deps. / sphalerite, siderite, hydrozincite, smithsonite, galena, anglesite, calcite, dolomite / tabular, yellow barytes	WULFENITE $Pb[MoO_4]$ (Yellow Lead Ore) / Mežice, Yugoslavia / WULFENITE $Pb[MoO_4]$ (Yellow Lead Ore) / Los Lamentos, Mexico
174	orange, yellow, brown, whitish, grey, greenish / white	ada-mantine, greasy / transluc.	3.5 — 4 / 7.1	good / brittle, uneven	columnar, ventri-cular, barrel-shaped, needle-shaped, thick-tabular crystals; botryoidal, reni-form, earthy, crusty, encrusta-tions, coatings / hexagonal	in lead ore deps. / arsenic ores, galena, cerussite, mimete-site, barytes, limo-nite, descloizite & o. / pyromorphite, vanadinite, cf. No. 146	MIMETESITE $Pb_6[Cl (AsO_4)_3]$ (Mimetite) / Dry Gill Mine, Cumberland, England
175	orange, yellow, brown, green / yellow	ada-mantine, greasy / transluc.	3.5 / 7.2	good / brittle, con-choidal	barrel-shaped, ventricular, thick-tabular, grown up crystals; in semi-globular agg. / hexagonal	in lead ore deps., cf. No. 327 / — / —	Mimetesite variety KAMPYLITE Pb_6 $[Cl (As,P) O_4]_3$ (Phosphomime-tite) / Příbram, Czechoslovakia

Hardness 7 — 8

No.	Colour / Streak	Lustre / Transparency	Hardness / Specific Gravity	Cleavage / Fracture & o. / Phys. Props.	Common Form, Aggregates / Crystalline Syst.	Occurrence / Assoc. Minerals / Similar Minerals	Name & Chem. Formula / Origin of Specimen
176	orange, yellowish, red, brownish / white	vitreous to greasy / transpar., transluc.	7 — 7.5 / 4.2	— / brittle, con-choidal to splintery	icositetrahedral, almost globular crystals; compact, massive / cubic	in granites, pegma-tites, in crystalline schists / — / other garnets, sphalerite, zircon	SPESSAR-TITE $Mn_3Al_2 [SiO_4]_3$ (Garnet Family) / Takovaja River, Central Urals, USSR

No.	Colour / Streak	Lustre / Transparency	Hardness / Specific Gravity	Cleavage / Fracture & o. / Phys. Props.	Common Form, Aggregates / Crystalline Syst.	Occurrence / Assoc. Minerals / Similar Minerals	Name & Chem. Formula / Origin of Specimen
177	copper-red, brown to black, weathered crust	metallic opaque	2.5 – 3 / 8.5 – 9	— / very ductile, malleable, sectile, flexible, hackly	cube-shaped, octahedral, tabular, tabular crystals; strongly distorted, dendritic, platy, filiform, mossy, arborescent, ramifying, compact, in clumps, sheets, plates, encrustations / cubic	in basalt & porphyritic rocks, in limestone veins / cuprite, chalcocite, malachite, azurite, calcite, analcime, limonite, quartz & o. / —	nat. COPPER Cu / Houghton Co., Michigan, USA
178	copper-red						nat. COPPER Cu / Moldova, Banat Romania
179	reddish, reddish-white, silver-grey / black	metallic opaque	5 / 8	— / brittle	small, thin-tabular crystals; mostly compact, tubular, finely granular, fibrous / tetragonal	in silver-nickel-cobalt ore veins / niccolite, chloanthite, bornite, chalcocite / niccolite, rammelsbergite	MAUCHERITE Ni_3As_2 / Eisleben, Harz, Germany
180	reddish, light copper-red, grey, brownish brown-black	metallic opaque	5.5 / 7.7 – 7.8	— / brittle, conchoidal	mostly compact, disseminated, botryoidal, reniform, granular, massive, knitted / hexagonal	in silver-nickel-cobalt ore veins / chloanthite, bismuth, silver, arsenic, rammelsbergite, safflorite, annabergite, barytes / maucherite	NICCOLITE NiAs (Arsenical Nickel) / Schladming, Styria, Austria

RED MINERALS OF NON-METALLIC LUSTRE

Hardness 1 – 2.5

No.	Colour / Streak	Lustre / Transparency	Hardness / Specific Gravity	Cleavage / Fracture & o. / Phys. Props.	Common Form, Aggregates / Crystalline Syst.	Occurrence / Assoc. Minerals / Similar Minerals	Name & Chem. Formula / Origin of Specimen
181	red, reddish-orange / yellow-orange	adamantine, greasy / transpar., transluc.	1.5 – 2 / 3.5 – 3.6	good / soft to brittle, sectile, decomposes in the light	short- to long-columnar, needle-shaped crystals; in druses; compact, granular, massive, disseminated, as encrustations / monoclinic	in ore veins, in marls, limestone, dolomite, in volcanoes & at hot springs / orpiment, arsenic & lead ores, antimonite / cinnabarite, crocoite	REALGAR As_4S_4 / Baia Sprie (Felsöbánya), Romania
182	cherry-red, purple-red / cherry-red, brown-red	adamantine / transluc.	1 – 1.5 / 4.6	very good / flexible	capillary, needle-shaped crystals; radial, bunchy, radiate-fibrous agg., disseminated / triclinic	in ore veins / antimonite, berthierite, valentinite, quartz / —	KERMESITE Sb_2S_2O (Red Antimony, Pyrantimonite) / Bräunsdorf, Germany
183	deep red orange, light yellow	vitreous / transpar., transluc.	1.5 – 2 / 2 – 2.1	very good / brittle	short-columnar crystals; granular, compact, botryoidal, bundle-shaped agg. / monoclinic	in ore veins in weathering zones / — / realgar, cinnabarite	BOTRYOGEN MgFe [OH (SO_4)_2] . 7H_2O / Redington Mine California, USA
184	scarlet-red, grey to black / scarlet-red	adamantine to dull / transpar., transluc., opaque	2 – 2.5 / 8.1	good / splintery, uneven	thin- to thick-tabular, cube-shaped, short-columnar, polymorphous crystals; compact, granular, massive, fibrous, earthy, disseminated / trigonal	in hydrothermal veins, as impregnation in sedimentary rocks, at thermal springs / quicksilver, chalcopyrite, pyrites, galena, carbonates, opal / realgar, rutile, crocoite, proustite, botryogen	CINNABARITE HgS (Cinnabar) / Tržić, Slovenia, Yugoslavia

178
180
182
184

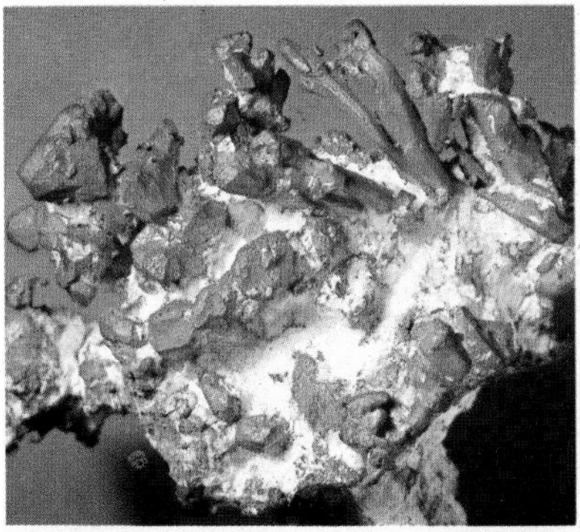

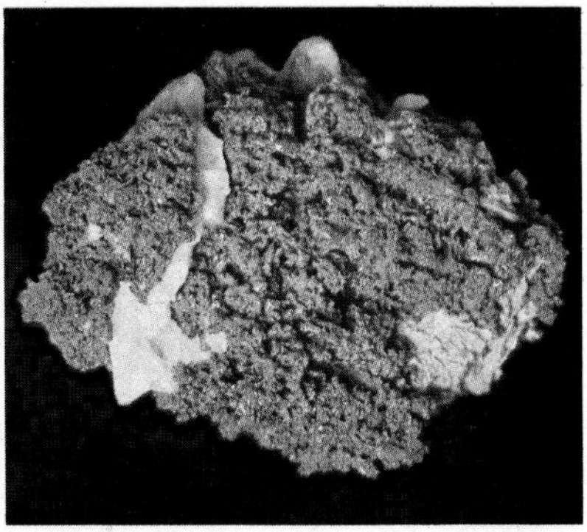

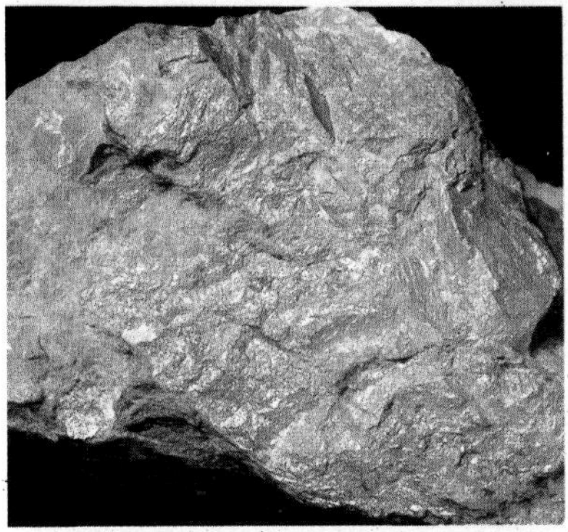

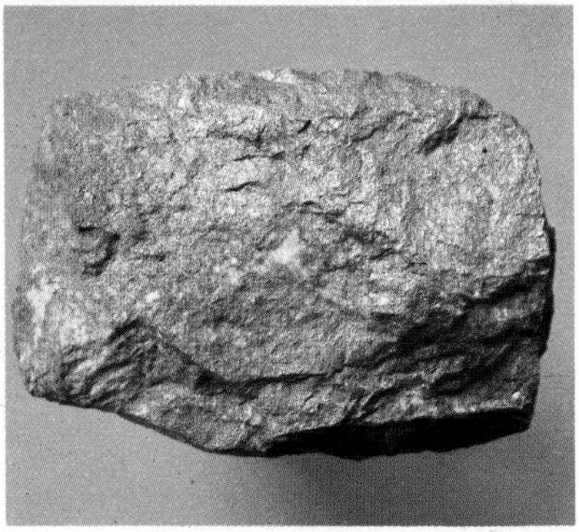

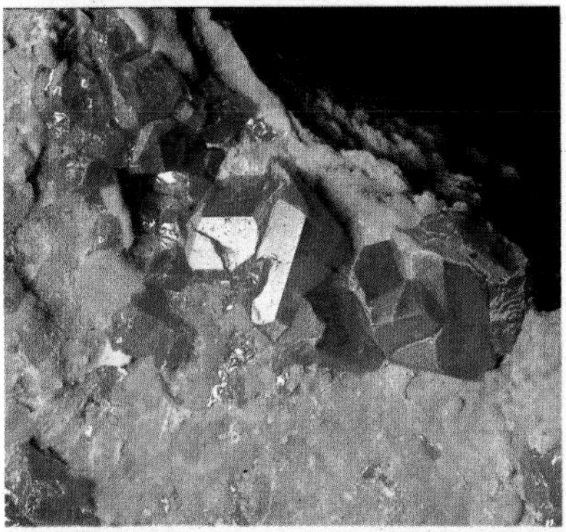

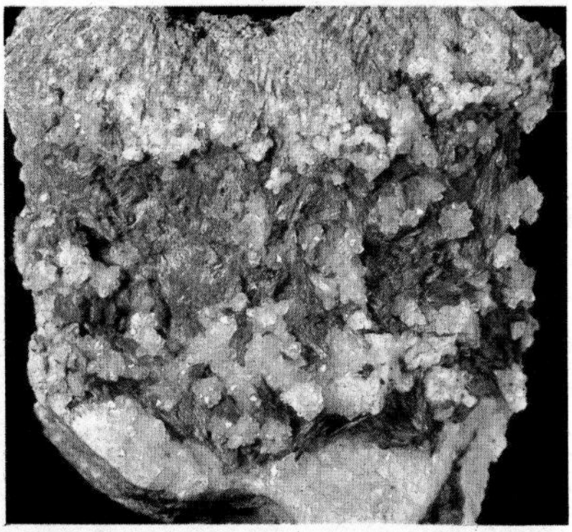

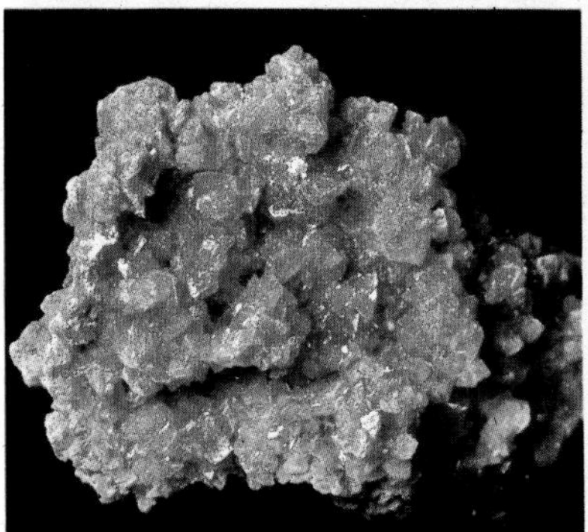

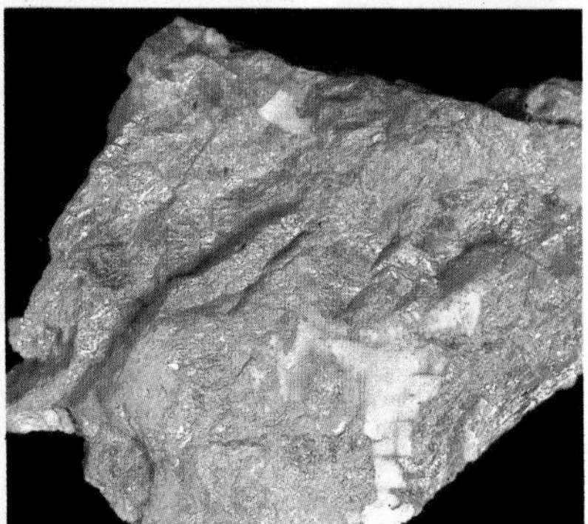

No.	Colour / Streak	Lustre / Transparency	Hardness / Specific Gravity	Cleavage Fracture & o. Phys. Props.	Common Form, Aggregates / Crystalline Syst.	Occurrence Assoc. Minerals / Similar Minerals	Name & Chem. Formula / Origin of Specimen
185	cinnabar-red, lead-grey / brick-red, yellowish-red	adamantine, transpar., transluc.	2.5 / 5.5	good splintery, conchoidal	prismatic, fragile, columnar, rhombo-hedral crystals; compact, massive, dendritic, as coatings, encrustations / trigonal	in silver-lead ore veins / silver, argentite, arsenic, chloanthite, galena, pyrites, barytes, fluorspar, pyrargyrite, cinnabarite	PROUSTITE Ag_3AsS_3 (Light Red Silver Ore) / Jáchymov, Czechoslovakia
186	dark red, reddish-lead-grey, black / dark red, brown-red	adamantine, metallic, dull transluc.	2.5−3 / 5.8	very good splintery, conchoidal	prismatic, thick-tabular, fragile, needle-shaped to capillary crystals; compact, granular, disseminated / trigonal	in silver-lead ore veins / proustite / proustite, cuprite, cinnabarite	PYRAR-GYRITE Ag_3SbS_3 (Dark Red Silver Ore) / St. Andreasber Harz, Germany
187	brick-red / orange-yellow	greasy, dull opaque	2−3 / 8.9−9.2	— earthy, uneven	only compact, fibrous, finely granular, earthy, disseminated, encrustations / tetragonal	weathering product of galena, cerussite & o. lead minerals / — / cinnabarite, realgar	MINIUM Pb_3O_4 (Red Oxide of Lead) / Niedermendig, Eifel, Germany
188	red, yellowish, red-orange / yellow, orange-yellow	adamantine, greasy transpar., transluc.	2.5−3 / 5.9−6	good soft, uneven	long-columnar to needle-shaped crystals; sometimes hollow showing longitudinal striation; in druses; also compact / monoclinic	in oxidation zone of ore veins / galena, limonite, quartz / realgar, cinnabarite	CROCOITE $Pb[CrO_4]$ (Red Lead Ore) / Dundas, Tasmania, Australia
189	red, red-brown / whitish	vitreous, pearly transpar., transluc.	2.5 / 2.8−3.2	very good soft to brittle	pseudohexagonal thin- to thick-tabular, embedded crystals / monoclinic	product of decomposition of biotite, as insets in basaltic rocks & tuffs / — / —	Biotite variety RUBELLAN $K(Mg,Fe,Mn)_3$ $[(OH,F)_2 Al Si_3 O_{10}]$ / Laacher See, Eifel, Germany

Hardness 3−4

No.	Colour / Streak	Lustre / Transparency	Hardness / Specific Gravity	Cleavage Fracture & o. Phys. Props.	Common Form, Aggregates / Crystalline Syst.	Occurrence Assoc. Minerals / Similar Minerals	Name & Chem. Formula / Origin of Specimen
190	reddish, violet, bluish, white, grey, colourless / white	vitreous, pearly transpar., transluc.	3−3.5 / 2.8−3	very good brittle, distinct cleavage cracks	tabular, short-columnar, cube-shaped crystals; spathic, fibrous, compact, botryoidal, granular, massive / orthorhombic	in ore veins, in salt deps. / halite, gypsum, dolomite / cryolite, gypsum, barytes, calcite, cf. No. 58	ANHYDRITE $Ca[SO_4]$ / Hallein near Salzburg, Austria
191	red-yellow, brown, green, black, colourless / white, yellow, brown	adamantine, greasy, almost metallic transpar. to opaque	3.5−4 / 3.9−4.2	very good, rhombo-dodecahedron brittle	tetrahedral, dodecahedral, cubic crystals; coarse to fine-grained, massive, compact, fibrous / cubic	in ore veins / galena, chalcopyrite & o. / cf. No. 142, 315, 445	SPHALERITE ZnS (Zinc Blende) / Cumberland, England
192	rose-red, raspberry-red, grey to brown / white	vitreous transluc.	3.5−4 / 3.3−3.6	very good, rhombo-hedron brittle	lentiform, rhombo-hedral crystals; often recurved, saddle-shaped agg., compact, granular, spathic, massive, botryoidal / trigonal	in ore veins, in iron ore deps., in limestones & argillaceous slates / haematite, limonite / rhodonite, quartz	RHODO-CHROSITE $MnCO_3$ (Manganese Spar) / Săcărâmb, Romania

No.	Colour / Streak	Lustre Transparency	Hardness / Specific Gravity	Cleavage / Fracture & o. / Phys. Props.	Common Form, Aggregates / Crystalline Syst.	Occurrence / Assoc. Minerals / Similar Minerals	Name & Chem. Formula / Origin of Specimen
193	rose-red, pink, reddish, yellowish, white / white	vitreous, pearly, transluc.	3.5 – 4 / 2.7	good / brittle	rhombohedral, lentiform, recurved crystals; globular, botryoidal agg.; compact / trigonal	manganiferous mixed crystals of calcite family, in ore veins / — / —	**MANGAN-CALCITE** $CaCO_3 \pm MnC($ / Banská Štiavnica, Czechoslovakia
194	brick-red, light pink / white	pearly, vitreous, dull transluc.	3.5 – 4 / 2.1 – 2.2	very good / brittle	tabular crystals in druses; rose-shaped agg. / monoclinic	in vesicles & cracks in basalt rocks, in ore deps. / calcite, quartz, other zeolites / cf. No. 71	**HEULANDIT** $Ca[Al_2Si_7O_{18}]$. $6H_2O$ (Stellerite, Zeolite Family) / Kilpatrick near Dumbarton, Scotland
195	reddish-lead-grey, red-brown, cochineal-red / brownish-red	ada-mantine, sub-metallic opaque	4 / 6.1	clear / brittle, uneven	octahedral, cubic, dodecahedral crystals; compact, massive, granular, earthy, dissemi-nated / cubic	as oxidation pro-duct of copper ores / copper, malachite, azurite, limonite / cinnabarite, proustite, haematite	**CUPRITE** Cu_2O (Red Oxides of Copper) / Bisbee, Arizona, USA
196	luminous-red, cochineal-red / brownish-red	ada-mantine transluc.	4 / 6.1	good / —	fibrous, capillary, acicular crystals; bunchy, interlacing / cubic	needle-shaped cuprite / — / —	Cuprite variety **CHALCO-TRICHITE** Cu_2O (Copper Protoxide) / Huel Basset, Cornwall, England

No.	Colour / Streak	Lustre Transparency	Hardness / Specific Gravity	Cleavage / Fracture & o. / Phys. Props.	Common Form, Aggregates / Crystalline Syst.	Occurrence / Assoc. Minerals / Similar Minerals	Name & Chem. Formula / Origin of Specimen
197	dark red, reddish-yellow, red-brown / orange-yellow, brownish-yellow	ada-mantine, sub-metallic transluc., opaque	4.5 – 5 / 5.4 – 5.7	very good / brittle	mostly granular, compact, spathic, lamellar / hexagonal	in metamorphic limestones / franklinite, calcite, willemite, rhodonite, garnet / rutile, cinnabarite	**ZINCITE** ZnO (Red Oxide of Zinc) / Franklin, New Jersey, USA
198	yellow, red-brown, brown to black / yellow, brownish-yellow	metallic, ada-mantine to dull transpar., transluc.	5.5 – 6 / 3.9 – 4.2	poor / subcon-choidal	flat, thin-tabular, pyramidal, only grown up crystals / orthorhombic	in alpine cracks / quartz, anatase, adular, titanite & o. / rutile, ilmenite	**BROOKITE** TiO_2 / Snowdon, North Wales, England
199	red, pink, white, yellowish / white	vitreous transpar., transluc.	5 – 5.5 / 2.2 – 2.4	very good / brittle	short-columnar, needle-shaped, capillary crystals in druses / orthorhombic	in cavities in igneous rocks / other zeolites, cal-cite, apophyllite & o. / cf. No. 88	**NATROLITE** $Na_2[Al_2Si_3O_{10}]$. $2H_2O$ (Zeolite Family) / Ústí-on-Elbe, Czechoslovakia
200	flesh-red, rose-red, raspberry-red, reddish-brown / whitish	vitreous, pearly transluc.	5.5 – 6.5 / 3.4 – 3.6	good / uneven	columnar, thick-tabular to needle-shaped crystals; compact, granular, coarse-spathic, massive, often stained black / triclinic	in manganese ore deps., in lead-zinc ore veins / magnetite, haus-mannite & o. / dialogite	**RHODONITE** $CaMn_4[Si_5O_{15}]$ (Manganese Spar, Fowlerite) / Franklin, New Jersey, USA

3	194
5	196
7	198
9	200

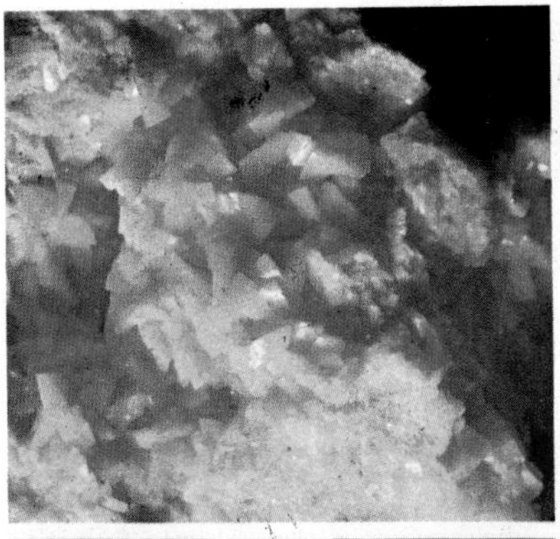

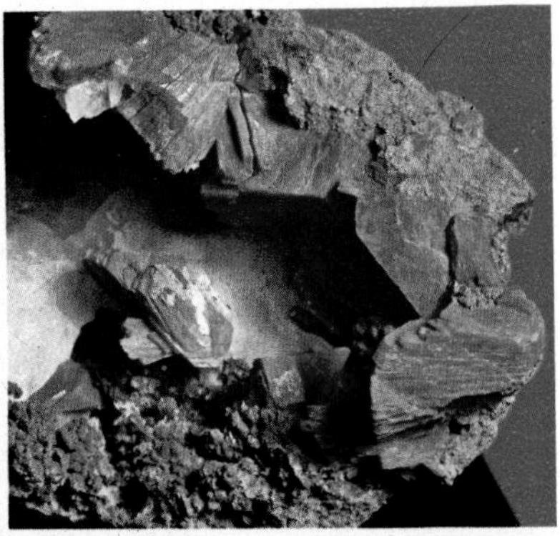

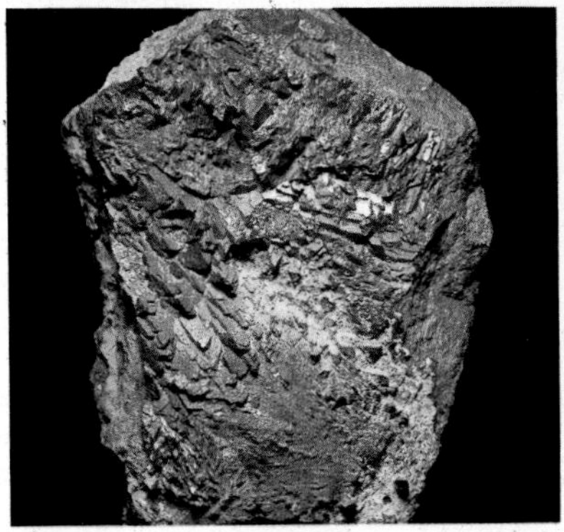

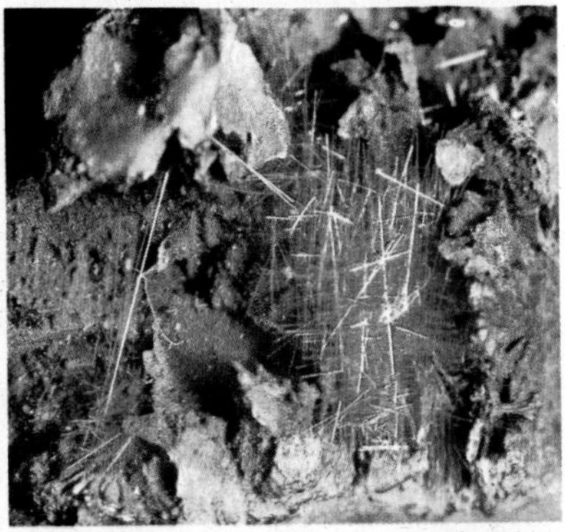

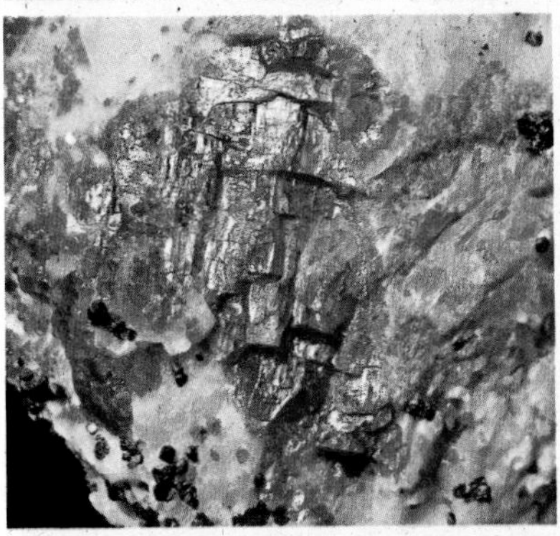

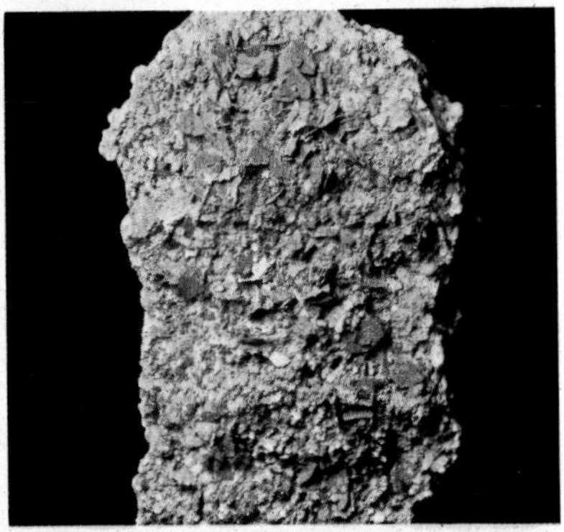

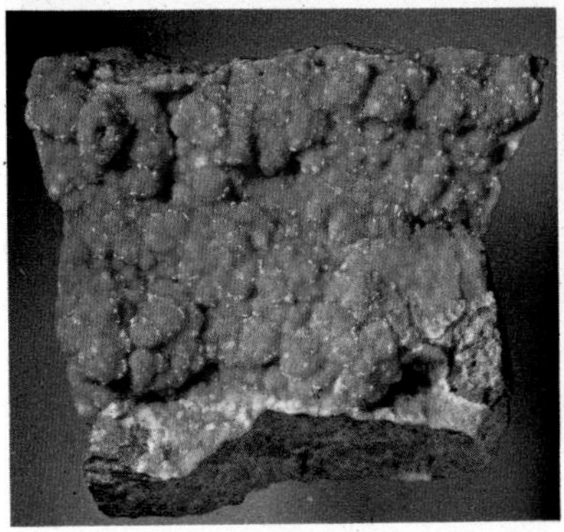

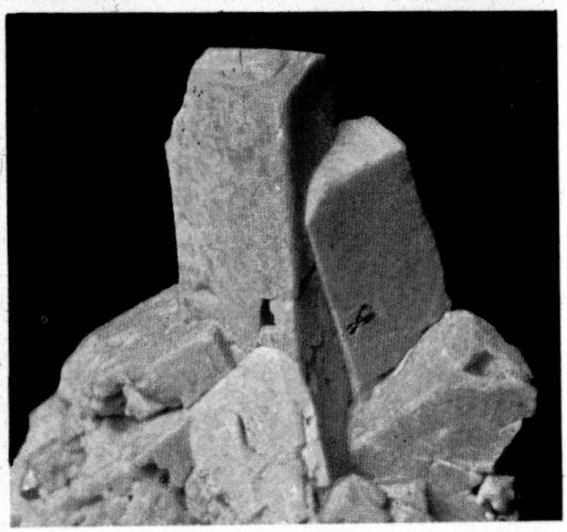

No.	Colour / Streak	Lustre / Transparency	Hardness / Specific Gravity	Cleavage / Fracture & o. / Phys. Props.	Common Form, Aggregates / Crystalline Syst.	Occurrence / Assoc. Minerals / Similar Minerals	Name & Chem. Formula / Origin of Specimen
201	red-brown, grey, black / blood-red, red-brown	sub-metallic, dull / opaque	5 — 6.5 / 5.2 — 5.3 / — / scaly slacking, fibrous	compact, massive, fibrous, haematitic, reniform, stalactitic, with fibrous fracture / trigonal	in ore veins, in various rocks / cinnabarite, proustite, cuprite, cf. No. 435 — 437 / —	Haematite variety **RED HAEMATITE** F_2O_3 / Blatno, Czechoslovakia
202	brick-red / white	vitreous / transpar., transluc.	6 / 2.1 / — / conchoidal	veins-nodules, fillings of cavities, compact, botryoidal, reniform / amorphous	in fissures & cracks in porphyritic trachyte / —	Opal variety **FIRE-OPAL** $SiO_2 . nH_2O$ / Zimapan, Mexico
203	yellow, red, brown-red, brownish-black / yellowish-brown, brown-red	ada-mantine, sub-metallic, greasy / transluc., opaque	6 — 6.5 / 4.2 — 4.3 / good uneven, conchoidal	thick-columnar, needle-shaped crystals; often heart-shaped & reniform twins; longitudinal striation, compact, granular, tubular, disseminated, detached grains / tetragonal	in pegmatites, in alpine cracks / titanite, anatase, brookite, felspar, apatite, quartz ilmenite, zincite, cassiterite, zircon, tourmaline	**RUTILE** TiO_2 / Alto Adige, Italy
204	reddish, grey-white, white, colourless / white	vitreous, pearly / transpar., transluc.	6 / 2.7 / very good brittle	tabular, short-columnar grown up & embedded crystals; massive, granular / triclinic	rock constituent in diff. rocks, in ore deps. / — / other felspars, cf. No. 102	**ANORTHITE** $Ca[Al_2Si_2O_8]$ (Felspar Family) / Toal della Foja, Trentino, Italy
205	reddish, brownish, grey, white, yellowish, colourless / white	vitreous, pearly / transluc., transpar.	6 — 6.5 / 2.5 / very good brittle, uneven	short-columnar, bench-shaped, in thick plates embedded & grown up crystals; often twinned, spathic masses, finely granular, massive / monoclinic	main constituent of many rocks, in granite druses, in pegmatites, in alpine cracks & crystalline schists, in ore veins / other felspars, spodumene cf. No. 346	**ORTHO-CLASE** $K[AlSi_3O_8]$ (Felspar Family) / Strigom, Poland
					Hardness 6 — 7.5		
206	reddish, brown, grey, white, green / white	vitreous, silky / transluc.	6 — 7 / 3.2 / very good brittle	columnar, needle-shaped crystals; fine-fibrous, wisp-like agg.; radial, tubular, compact / orthorhombic	in crystalline schists, granulites, eclogites, in contact-metamorphic rocks / — / cyanite	**SILLI-MANITE** $Al_2[OSiO_4]$ (Fibrolite) / Goyamin Pool, Australia
207	red, reddish, brown / white, reddish	vitreous / opaque	7 / 2.6 / — / brittle, conchoidal	columnar, grown up crystals with druse-shaped haematite coating / trigonal	quartz coloured with iron oxide / — / —	Quartz variety **EISENKIESEL** SiO_2 (Ferruginous Quartz) / Cínovec, Czechoslovakia
208	red, pink, violet / white	vitreous, greasy / transpar., transluc.	7 — 7.5 / 3 — 3.2 / — / uneven	columnar to needle-shaped crystals; radiating agg.; tubular, compact / trigonal	in pegmatites, granites / lepidolite, indigolite, quartz & o. cf. No. 220	**RUBELLITE** Na,Li,Mg_3Al_6 $[(OH)_3(BO_3)_3$ $Si_6O_{18}]$ / Rožná, Czechoslovakia

VIOLET MINERALS OF NON-METALLIC LUSTRE

No.	Colour Streak	Lustre Transparency	Hardness Specific Gravity	Cleavage Fracture & o. Phys. Props.	Common Form, Aggregates Crystalline Syst.	Occurrence Assoc. Minerals Similar Minerals	Name & Chem. Formula Origin of Specimen
209	blue, violet, greenish, black colourless, changing immediately to indigo blue	vitreous, pearly, sub-metallic transluc., opaque	1.5—2 2.6—2.7	very good soft, thin laminae, flexible	long-columnar, needle-shaped, tabular, chip-shaped crystals; compact, reniform, globular, radial, earthy, powdery monoclinic	in clays, peat, in fossil shells limonite lazulite	**VIVIANITE** $Fe_3[PO_4]_2 . 8H_2O$ (Blue Iron Earth) Bodenmais, Bavarian Forest, Germany
210	reddish, pink, violet, grey reddish, rose-red	adamantine, pearly transluc.	1.5—2.5 3.0	very good soft	needle-shaped, scaly, lamellar crystals; in bunchy, stellate clusters; earthy encrustations monoclinic	decomposition product of cobalt ores pharmacolite & o. —	**ERYTHRITE** $Co_3(AsO_4)_2 . 8H_2O$ (Cobalt Bloom) Schneeberg, Ore Mountains, Germany

Hardness 2.5—4

No.	Colour Streak	Lustre Transparency	Hardness Specific Gravity	Cleavage Fracture & o. Phys. Props.	Common Form, Aggregates Crystalline Syst.	Occurrence Assoc. Minerals Similar Minerals	Name & Chem. Formula Origin of Specimen
211	violet, pale red, reddish-grey, greenish, white whitish, light pink	pearly transpar., transluc.	2.5—4 2.8—2.9	very good soft, laminae elastic, flexible	lamellar, pseudo-hexagonal crystals; scaly, compact, finely granular monoclinic	in pegmatites & granites in tin ore deps. rubellite, topaz, beryllium, spodumene zinnwaldite	**LEPIDOLITE** $KLi_2Al[(OH,F)_2Si_4O_{10}]$ (Lithium Mica) Maharitra, Madagascar
212	violet, yellow, green, blue, pink, colourless white	vitreous transpar., transluc.	4 3.0—3.2	perfect conchoidal, brittle	cubic, octahedral crystals; coarsely to finely granular, spathic, massive, compact, striated cubic	in ore veins, in rock fissures — apatite, amethyst, cf. No. 73, 233	**FLUORITE** CaF_2 (Fluor Spar) Alston Moor, Westmorland, England

Hardness 4—7

No.	Colour Streak	Lustre Transparency	Hardness Specific Gravity	Cleavage Fracture & o. Phys. Props.	Common Form, Aggregates Crystalline Syst.	Occurrence Assoc. Minerals Similar Minerals	Name & Chem. Formula Origin of Specimen
213	colourless, white, reddish, violet, greenish white	vitreous, pearly transpar., transluc.	4—5 2.3—2.4	very good brittle, uneven	short-columnar, tabular, pyramidal crystals; compact, granular, lamellar, massive tetragonal	in druses in cavities in basaltic rocks, in ore veins, in magnetite ore deps. analcime, natrolite, calcite —	**APOPHYLLITE** $KCa_4[F(Si_4O_{10})_2] . 8H_2O$ St. Andreasberg, Harz, Germany
214	colourless, white, yellow, violet, green white	vitreous, greasy transpar., transluc.	5 3.1	clear brittle, conchoidal	hexagonal, short- to long-columnar crystals; compact, massive, granular hexagonal	as constituent in magmatic rocks, in ore veins — cf. No. 158, 281	**APATITE** $Ca_5[F(PO_4)_3]$ Slavkov, Czechoslovakia
215	dark violet, blue-violet white, grey	vitreous, greasy transluc., opaque	5—6 3.2	good —	foliaceous, granular-radiating, fine-fibrous masses monoclinic	violet or blue manganiferous variation of diopside — —	Diopside variety **VIOLANE** $CaMg[Si_2O_6] \pm Mn,Fe$ St. Marcel, Piedmont, Italy
216	red-violet, blue, greenish white	silky transluc.	7 3.4	clear —	in compact pieces, thin-columnar, needle-shaped, parallel-fibrous orthorhombic	in pegmatite veins, in gneiss, in granulites cyanite, cordierite cyanite	**DUMORTIERITE** $(Al,Fe)_7[O_3BO_3(SiO_4)_3]$ Yuma, Arizona, USA

No. 209—216

Table 2

9	210
1	212
3	214
5	216

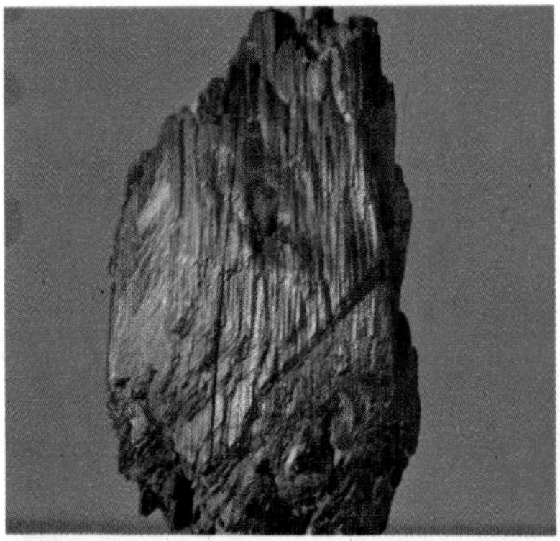

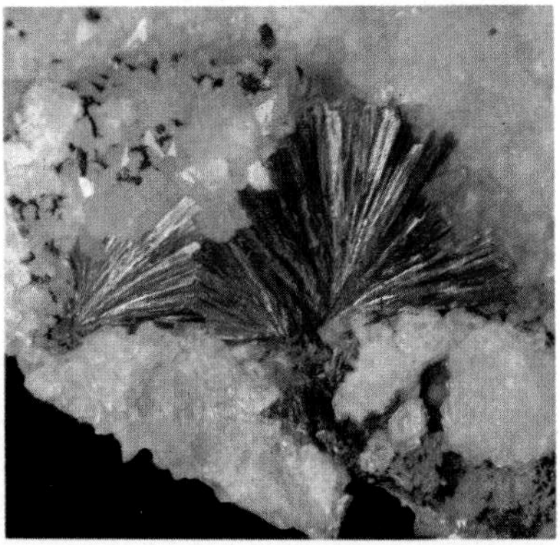

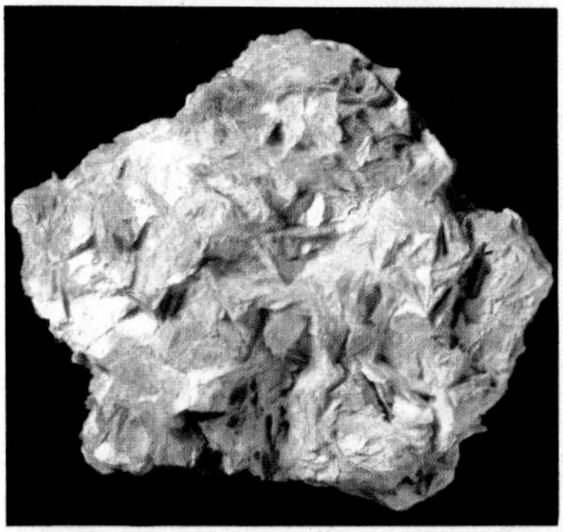

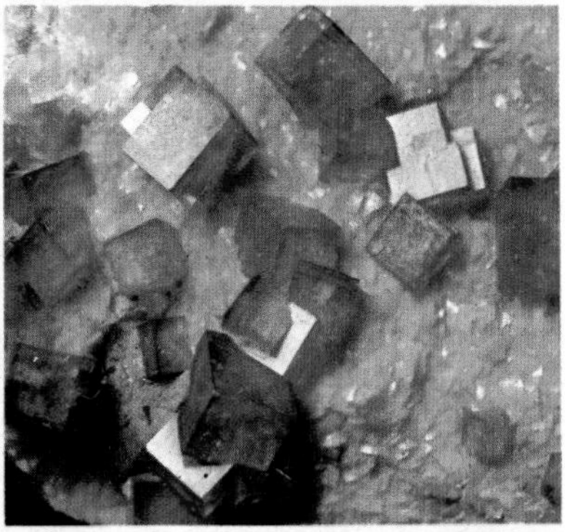

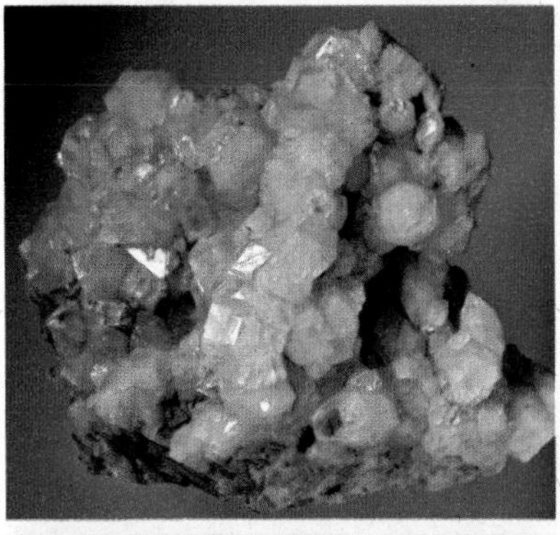

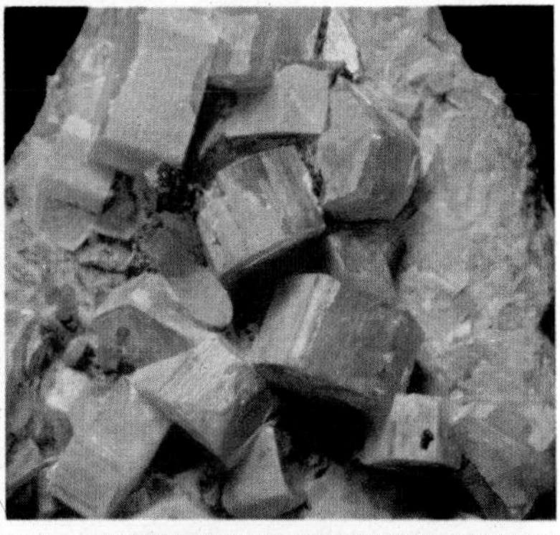

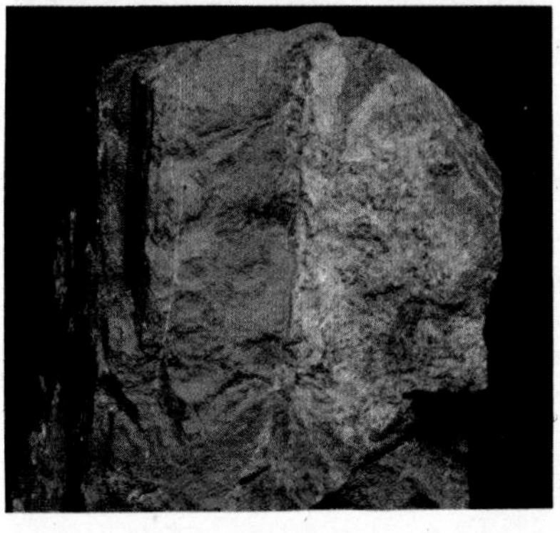

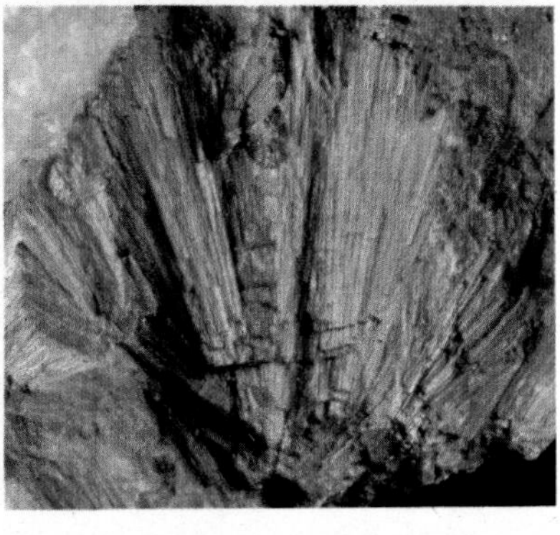

No.	Colour Streak	Lustre Transparency	Hardness Specific Gravity	Cleavage Fracture & o. Phys. Props.	Common Form, Aggregates Crystalline Syst.	Occurrence Assoc. Minerals Similar Minerals	Name & Chem. Formula Origin of Specimen
217	violet	vitreous, greasy	$\frac{7}{2,6}$	—	hexagonal, columnar, grown up crystals in druses	colour variety of quartz, in druses of volcanic effusive rocks, in ore veins	**AMETHYST** SiO_2 Matto Perro, Uruguay
218	white	transpar., transluc.		brittle, conchoidal	trigonal	— apatite, fluorite	**AMETHYST** SiO_2 Banská Štiavnica, Czechoslovakia
219	red, brown, violet to black white	vitreous, greasy transpar., opaque	$\frac{7}{4.1-4.3}$	— brittle, splintery, uneven	rhombododeca-hedral crystals; compact, granular, massive, dissemi-nated cubic	insets in gneiss & mica schists, in placers — cf. No. 526	**ALMANDINE** $Fe_3Al_2[SiO_4]_3$ (Garnet Family) Bodö, Nordland, Norway
220	violet, red, pink white	vitreous, greasy transpar., transluc.	$\frac{7-7.5}{3-3.2}$	— uneven	columnar to needle-shaped crystals; tubular, radial agg. trigonal	in pegmatites & granites — cf. No. 208	**RUBELLITE** $Na,Li,Mg_3,Al_6[(OH)_3(BO_3)_3Si_6O_{18}]$ (Tourmaline Family) Pala Chief, Calif., USA

BLUE MINERALS OF METALLIC LUSTRE   *Hardness 1.5 — 3*

No.	Colour Streak	Lustre Transparency	Hardness Specific Gravity	Cleavage Fracture & o. Phys. Props.	Common Form, Aggregates Crystalline Syst.	Occurrence Assoc. Minerals Similar Minerals	Name & Chem. Formula Origin of Specimen
221	blue, bluish-black bluish-black	metallic, greasy opaque	$\frac{1.5-2}{4.6}$	very good soft, flexible	thin-tabular, foliaceous crystals; mostly compact, finely foliaceous, massive, nodular, platy, powdery hexagonal	weathering product in copper ore veins, as volcanic subli-mation product chalcopyrite, chalcosine bornite	**COVELLITE** CuS (Indigo Copper) Alghero, Sardinia
222	grey, blue & tarnished brownish-black	metallic opaque	$\frac{2-3}{3.9-4.3}$	good brittle	needle-shaped, fibrous, capillary crystals; fibrous, tubular, massive agg. orthorhombic	in silver ore veins antimonite, arseno-pyrite, quartz antimonite	**BERTHIE-RITE** $FeS . Sb_2S_3$ Herja (Chiuz-baia, Kisbánya), Romania

BLUE MINERALS OF NON-METALLIC LUSTRE   *Hardness 1 — 2*

No.	Colour Streak	Lustre Transparency	Hardness Specific Gravity	Cleavage Fracture & o. Phys. Props.	Common Form, Aggregates Crystalline Syst.	Occurrence Assoc. Minerals Similar Minerals	Name & Chem. Formula Origin of Specimen
223	sky-blue, azure blue	silky, velvety transluc.	$\frac{1}{2.7}$	good —	fine-fibrous, capillary crystals; bunchy, globular agg.; velvety crusts orthorhombic	weathering product in ore veins malachite, azurite, limonite & o. aurichalcite	**CYANO-TRICHITE** $Cu_4Al_2[(OH)_{12}SO_4] . 2H_2O$ (Lettsomite) Moldova, Banat, Romania
224	colourless, yellow, red, blue, violet white	vitreous transpar., transluc.	$\frac{2}{2.1-2.2}$	perfect, cube soft to brittle, water-soluble	cube-shaped crystals; coarsely to finely granular, compact, spathic, fibrous, massive cubic	isolated deps. anhydrite, gypsum & o. — cf. No. 33	**ROCK SALT** $NaCl$ (Halite) Hallstatt, Salz-kammergut, Austria

7	218
9	220
1	222
3	224

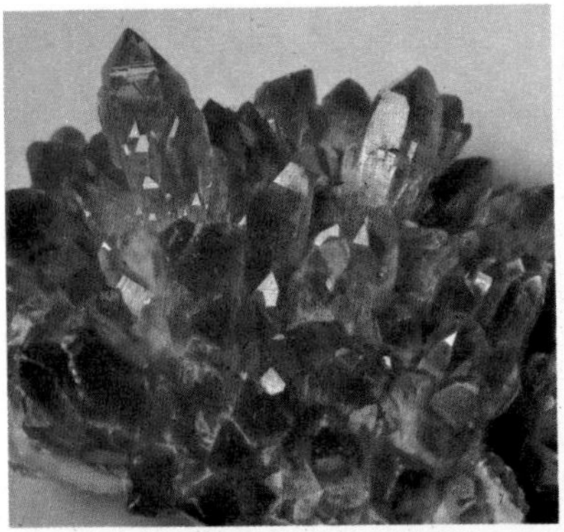

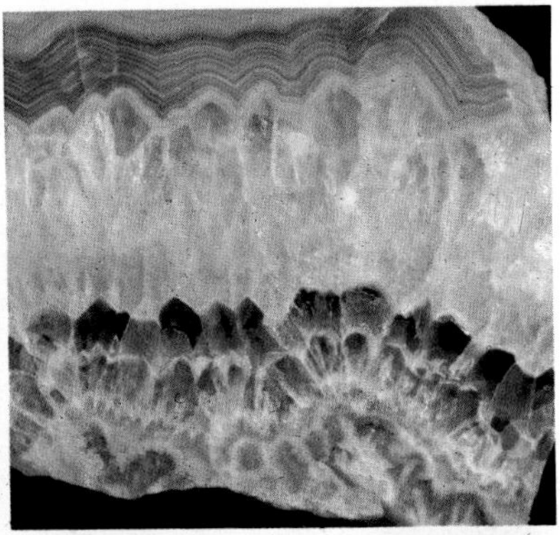

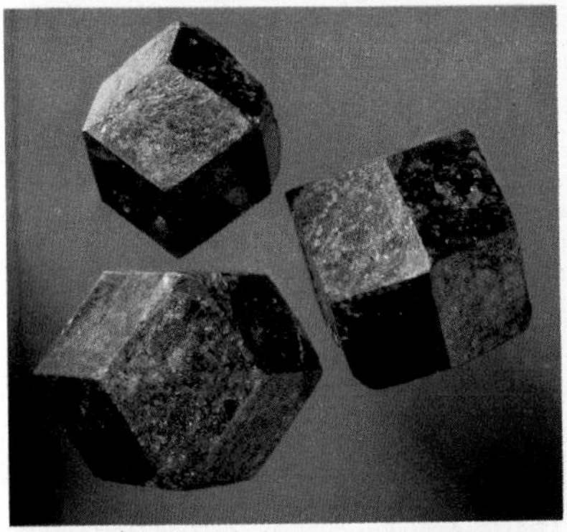

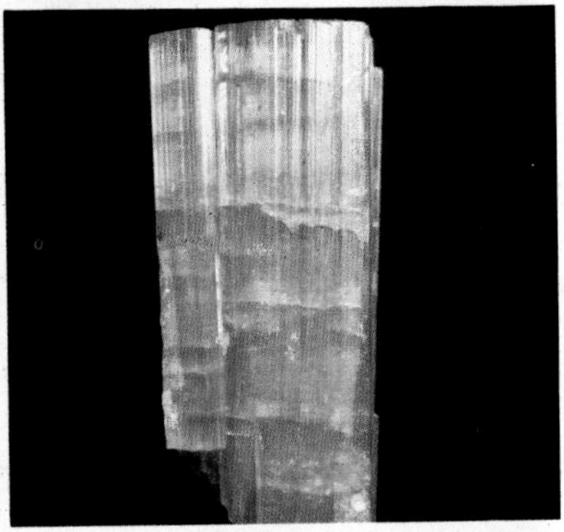

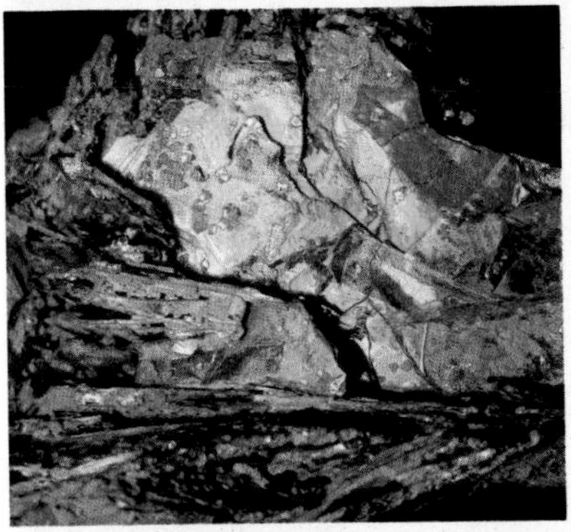

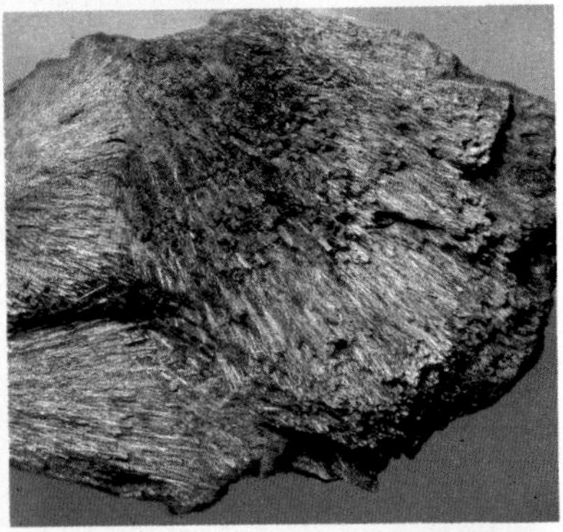

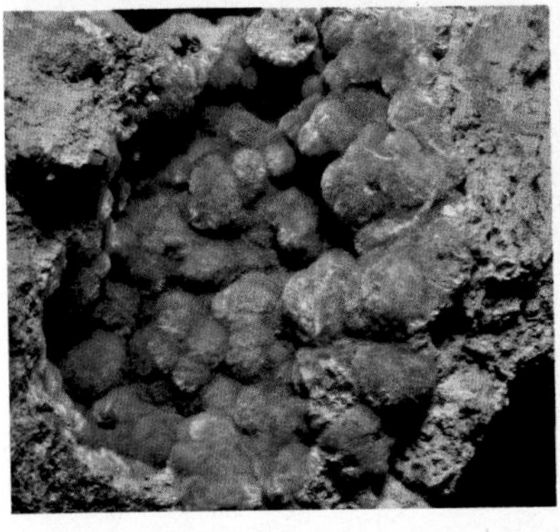

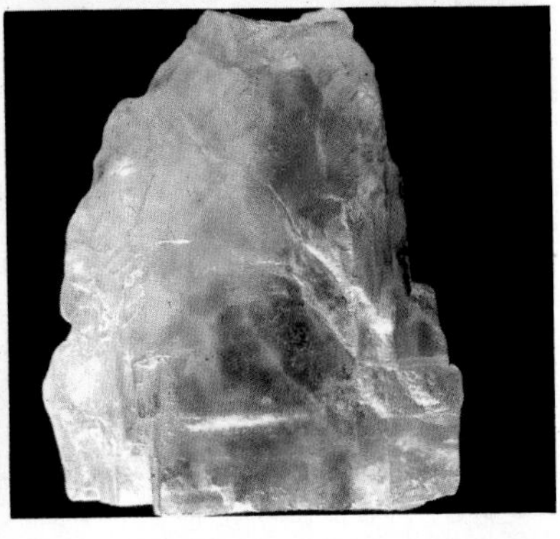

No.	Colour / Streak	Lustre / Transparency	Hardness / Specific Gravity	Cleavage / Fracture & o. / Phys. Props.	Common Form, Aggregates / Crystalline Syst.	Occurrence / Assoc. Minerals / Similar Minerals	Name & Chem. Formula / Origin of Specimen
225	light blue, greenish-blue / bluish, greenish	pearly, silky transluc.	2 / 3.6	very good / soft	fine needle-shaped, finely foliaceous crystals; fibrous, scaly, granular, compact, massive, as efflorescences / orthorhombic	in oxidation zone of copper-zinc veins / limonite, malachite, smithsonite / chrysocolla, cyanotrichite	AURI-CHALCITE $(Zn,Cu)_5$ $[(OH)_3 CO_3]_2$ / Zimapan, Mexico
226	sky-blue / light blue	ada-mantine, vitreous transluc.	2.5 / 5.3 — 5.5	very good / brittle to soft	short needle-shaped, tabular, grown up crystals; radial, fibrous, disseminated, crusty, encrustations / monoclinic	in oxidation zone of ore veins / chalcopyrite, galena, malachite, cerussite, barytes, quartz & o. / azurite, lazurite	LINARITE PbCu $[(OH)_2 SO_4]$ / Red Gill near Keswick, Cumberland, England
227	blue / white	vitreous transluc.	2.5 / 2.2 — 2.3	indistinct brittle, con-choidal, water-soluble	tabular, short-columnar crystals; crusty, reniform, fibrous, granular, scaly, massive, as efflorescences / triclinic	weathering product in oxidation zone of copper ores / — / liroconite, kröhn-kite	CHAL-CANTHITE $Cu[SO_4] . 5H_2O$ (Copper Vitriol) / Rio Tinto, Spain
228	blue-green, blue-grey / bluish-green	vitreous, greasy transluc.	2 — 2.5 / 2.9 — 3	indistinct brittle to soft	flat, lentiform, short-columnar crystals; in druses; compact, disseminated / monoclinic	as weathering product in ore veins / malachite, limonite, quartz & o. / chalcanthite	LIROCONITE $Cu_2Al[(OH)_4 AsO_4] . 4H_2O$ / Redruth, Cornwall, England
229	blue, greenish-blue / white	vitreous transpar., transluc.	2.5 — 3 / 2	very good water-soluble	tabular, long-co-lumnar, octahedral crystals; granular, fibrous, platy, crusts, fibrous agg. / monoclinic	weathering product in oxidation zone of copper ores / — / chalcanthite	KRÖHNKITE $Na_2[CuSO_4]_2 . 2H_2O$ / Chuquicamata, Chile
230					short-columnar, thick-tabular, needle-shaped crystals; compact, reniform, radial, massive, earthy, botryoidal, encrustations, coatings / monoclinic	in oxidation zone of copper ores, impregnation in sedimentary rocks / malachite, limonite, cuprite, chalco-pyrite & o. / linarite, lazurite	AZURITE $Cu_3[OH CO_3]_2$ (Blue Carbonate of Copper) / Tsumeb, South-West Africa
231	sky-blue, blackish-blue / blue	vitreous to dull transluc., opaque	3.5 / 3.7 — 3.8	good brittle, uneven			AZURITE $Cu_3[OH CO_3]_2$ (Blue Carbonate of Copper) / Chessy near Lyon, France
232	colourless, white, blue, reddish, yellowish / white	vitreous, pearly transpar., transluc.	3 — 3.5 / 3.9 — 4	good brittle, uneven, fibrous	thick-tabular, short-columnar crystals; granular, spathic, massive, tubular, fibrous, nodular, compact / orthorhombic	in fissures, in rock salt deps., in ore veins; as exhalation product, in cracks in limestones & dolomites / aragonite, calcite, sulphur, gypsum / barytes, cryolite, gypsum, calcite	COELESTINE $Sr[SO_4]$ / Špania Dolina, Czechoslovakia

No.	Colour / Streak	Lustre / Transparency	Hardness / Specific Gravity	Cleavage / Fracture & o. / Phys. Props.	Common Form, Aggregates / Crystalline Syst.	Occurrence / Assoc. Minerals / Similar Minerals	Name & Chem. Formula / Origin of Specimen
233	colourless, blue, pink, yellow, green, violet / white	vitreous transpar., transluc.	4 / 3—3.2	perfect, octahedron / brittle, conchoidal	cubic, octahedral crystals; granular, spathic, massive, compact / cubic	in ore veins, in rock cracks / — / barytes, apatite, amethyst, cf. No. 73, 212	FLUORITE CaF_2 (Fluor Spar) / Weisseck, Salzburg, Austria
234	colourless, white, yellow, blue, grey, red, brown / white	vitreous, pearly, greasy / transluc., transpar.	3—3.5 / 4.4	very good / less brittle, conchoidal	thin- to thick-tabular, columnar, lamellar crystals; massive, reniform, granular / orthorhombic	in ore veins, in cracks in sedimentary rocks / — / aragonite, calcite, coelestine, anhydrite, cf. No. 57, 144, 145	BARYTES $Ba[SO_4]$ (Heavy Spar) / Dědova Hora, Czechoslovakia
235	dark blue to blue-black, yellow, red / white, yellowish	adamantine, greasy, metallic / transluc., opaque	5.5—6 / 3.8—3.9	very good brittle	pointed-pyramidal, columnar, thick-tabular, only grown-up crystals / tetragonal	in alpine fissures, in cracks in crystalline schists / rutile, titanite, albite, quartz, chlorite, epidote / —	ANATASE TiO_2 / Sonnblick, Salzburg, Austria
236	colourless, white, blue, grey, greenish-grey, yellowish / white	vitreous, pearly / transpar., transluc.	in longitudinal direction 4, diagon. 6—7 / 3.6—3.7	very good brittle	wide-tubular, lamellar, flat-tabular crystals; diagonal striation, compact, radiating-lamellar to acicular masses / triclinic	in crystalline schists, granulites, in pegmatites / staurolite, andalusite, almandine, corundum, cf. No. 492	KYANITE $Al_2[O\,SiO_4]$ (Disthene) / Pizzo Forno, Tessin, Switzerland
237	colourless, white, blue, grey, greenish / white	vitreous, greasy / transpar., transluc.	5—6 / 2.3—2.5	good conchoidal	rarely isometric crystals; mostly compact, granular / cubic	constituent in igneous rocks / — / leucite, analcime, hauyne, nosean, lazurite, cf. No. 93	SODALITE $Na_8[Cl_2(AlSiO_4)_6]$ / Timmins, Ontario, Canada
238	blue, green-blue, green, red, yellow, grey, white / white, bluish	vitreous, greasy / transpar., transluc.	5.5—6 / 2.4—2.5	very good conchoidal	isometric, cubic crystals; compact, granular agg. / cubic	embedded in basaltic rocks; in volcanic ejecta / leucite, nepheline, & o. / sodalite, nosean, lazurite	HAÜYNITE (HAUYNE) $(Na,Ca)_{8-4}[(SO_4)_2(AlSiO_4)_6]$ / Niedermendig, Eifel, Germany
239	blue, bluish-green, white / white	vitreous transluc., opaque	6 / 3.1—3.2	— splintery, uneven	short-columnar, tabular, pyramidal crystals; mostly compact, massive, granular masses / monoclinic	in carbonate-quartz veins, in argillaceous slates, xenoliths in quartzites, in tin-pegmatite veins / quartz, muscovite & o. / lazurite, turquoise	LAZULITE $(Mg,Fe)Al_2[OH\,PO_4]_2$ (Blue Spar) / Werfen, Salzburg, Austria
240							LAZULITE $(Mg,Fe)Al_2[OH\,PO_4]_2$ (Blue Spar) / Sticklberg, Austria

No.	Colour / Streak	Lustre / Transparency	Hardness / Specific Gravity	Cleavage / Fracture & o. Phys. Props.	Common Form, Aggregates / Crystalline Syst.	Occurrence / Assoc. Minerals / Similar Minerals	Name & Chem. Formula / Origin of Specimen
241	blue, blue-grey / white	vitreous opaque, dull	6.5 / 3.6	— / conchoidal	idiomorphic, pyramidal, embedded crystals; no agg. / trigonal	embedded in granular natrolite / neptunite, anatase / —	**BENITOITE** $BaTi[Si_3O_9]$ / San Benito County, California, USA
242	blue / white	vitreous, greasy transluc.	6.5 / 3.2 − 3.4	— / splintery, uneven	short-columnar, pseudocubic, dipyramidal crystals; fine-grained to massive, compact / tetragonal	blue, cupreous vesuvianite, in metamorphic rocks / thulite / —	Vesuvianite variety **CYPRINE** $Ca_{10}(Mg,Fe_2)Al_4 [(OH,F)_4 (SiO_4)_5 (SiO_7)_2] \pm Cu$ / Souland, Norway
243	azure-blue, dark blue / light blue, greenish	pearly transluc.	3.5 / 4.8 − 5.1	very good / —	cube-shaped, pseudocubic crystals / tetragonal	as weathering product in copper deps. / malachite, azurite, cuprite, atacamite / —	**BOLEITE** $(26PbCl_2 . 3AgCl) . 24Cu(OH)_2 . 6AgCl . 3H_2O$ / Boleo, Mexico
244	blue, greenish-blue, grey, violet / white	vitreous, greasy transpar., transluc.	7 − 7.5 / 2.6	good / conchoidal, strongly pleochroic	short-columnar, platy crystals; mostly compact, granular, embedded / orthorhombic	in pegmatites, in metamorphic rocks / — / nepheline, quartz	**CORDIERITE** $Mg_2[Al_4Si_5O_{18}]$ (Dichroite, Iolite) / Dolní Bory, Czechoslovakia
245	blue / white	vitreous, greasy, dull opaque, transluc.	7 − 7.5 / 3 − 3.2	— / fissile, uneven, splintery	long-columnar to needle-shaped crystals; longitudinally striated; tubular to fibrous agg.; massive / trigonal	in pegmatites, in granites / lepidolite, rubellite / —	Tourmaline variety **INDIGOLITE** $(Na,Ca) (Mg,Al,Li)_3 (Al,Fe,Mg)_6 [(OH)_3 (BO_3)_3 Si_6O_{18}]$ / Rožná, Czechoslovakia
246	blue, red, yellow, brown, black / white	vitreous transpar., transluc.	8 / 3.5 − 4.1	indistinct / conchoidal	octahedral, embedded crystals; granular / cubic	in metamorphic limestones & dolomites, in basalts / corundum, zircon, garnet, cf. No. 360	**SPINEL** Al_2MgO_4 / Åker, Södermanland, Sweden

GREEN MINERALS OF NON-METALLIC LUSTRE *Hardness 1 − 2*

No.	Colour / Streak	Lustre / Transparency	Hardness / Specific Gravity	Cleavage / Fracture & o. Phys. Props.	Common Form, Aggregates / Crystalline Syst.	Occurrence / Assoc. Minerals / Similar Minerals	Name & Chem. Formula / Origin of Specimen
247	green, brownish-green, black-green / green, blue-green	dull opaque	1 − 2 / 2.8 − 2.9	— / —	massive, compact, earthy, powdery, loose / monoclinic	in vesicles in basaltic tuffs, in sedimentary rocks, in marls, limestone / calcite, zeolite / glauconite	**SELADONITE** $K(Fe,Mg) [(OH)_2AlSi_3O_{10}]$ (Green Earth) / Kadaň, Czechoslovakia
248	blackish-green, bluish-green / white	vitreous, greasy, pearly transluc., opaque	2 / 2.5 − 2.7	very good / soft, inelastic, flexible	thick-tabular, columnar crystals; often distorted; flakes; lamellar agg.; as dusty crusts, earthy, powdery / monoclinic	constituent in crystalline schists; in alpine cracks, as weathering product / — / green mica	**CLINO-CHLORE** $(Mg,Al)_3 [(OH)_2 AlSi_3O_{10}]Mg_3 (OH)_6$ (Chlorite Family) / Zillertal, Tyrol, Austria

No.	Colour / Streak	Lustre / Transparency	Hardness / Specific Gravity	Cleavage Fracture & o. Phys. Props.	Common Form, Aggregates / Crystalline Syst.	Occurrence Assoc. Minerals / Similar Minerals	Name & Chem. Formula / Origin of Specimen
249	green, greenish-white / pale green	dull transluc., opaque	$1-2.5$ / $3-3.1$	good soft	needle-shaped, capillary, lamellar crystals; mostly earthy, powdery, reniform, scaly, encrustations & coatings, efflorescences / monoclinic	as weathering product in nickel-cobalt ores / erythrite & o. / —	**ANNABERGITE** $Ni_3(AsO_4)_2 . 8H_2O$ (Nickel Bloom) / Jáchymov, Czechoslovakia
250	green, light blue / bluish-green	pearly, vitreous transpar., transluc.	$1.5-2$ / 3.2	very good soft, elastic, flexible	lamellar crystals; mostly radiating, flabelliform, scaly agg.; compact, reniform, spumous / orthorhombic	weathering product of copper ores / malachite, azurite / —	**TYROLITE** $Ca_2Cu_9[(OH)_{10}(AsO_4)_4] . 10H_2O$ / Schwaz, Tyrol, Austria
251	green / green-yellow	pearly, dull transluc.	$1-2$ / 3.5	good / —	needle-shaped, thick-lamellar crystals; radiating agg.; crusty, earthy, powdery, as coatings / monoclinic	decomposition product of uranium ores / other secondary uranium minerals / —	**CUPROSKLODOWSKITE** $CuH_2[UO_2SiO_4]_2 . 5H_2O$ (Jachimovite) / Jáchymov, Czechoslovakia
252	green, yellowish / white	vitreous transpar., transluc.	2 / 1.9	very good conchoidal, brittle, water-soluble	columnar, tabular, needle-shaped crystals; platy, stalactitic agg.; crusts, efflorescences / monoclinic	decomposition product of pyrites, pyrrhotine, in coal seams, in alum schists, marls / — / goslarite	**MELANTERITE** $Fe[SO_4] . 7H_2O$ (Copperas, Green Vitriol) / Kaznějov, Czechoslovakia
253	emerald-green / light green	pearly, vitreous transpar., transluc.	2 / $2.4-2.6$	very good brittle	thin-tabular crystals; in druses, also lamellar agg.; massive / trigonal	weathering product in ore veins / malachite, azurite, chalcopyrite, cuprite, limonite / —	**CHALCOPHYLLITE** $(Cu,Al)_3[(OH)_4(AsO_4,SO_4)] . 6H_2O$ / Redruth, Cornwall, England
254	green / pale green	vitreous, pearly, dull transpar., transluc.	$1-2$ / $2.9-3.1$	very good soft	thin-tabular, platy crystals; mostly fine-grained, compact, as coatings, encrustations, radiating fibrous agg.; disseminated / monoclinic	weathering product in ore veins, in dolomitic limestones / adamite / —	Annabergite variety **CABRERITE** $(Ni,Mg)_3[AsO_4]_2 . 8H_2O$ / Laurion, Greece
255	green, brownish, brown-yellow, greenish-white / white	silky transluc.	2 / $2.9-3$	good fibrous structure, soft	short- to long-parallel, fibrous, capillary agg. / monoclinic	filling in cracks & veins in serpentinites, from hydrothermally decomposed olivine, pyroxene, amphibole & o. / chromite, garnierite, pyrope, magnesite, talc, opal / —	Chrysotile variety **ASBESTOS** $Mg_6[(OH)_8Si_4O_{10}]$ / Alto Adige, Italy
256							Chrysotile variety **PICROSMINE** $Mg_6[(OH)_8Si_4O_{10}]$ / Alto Adige, Italy

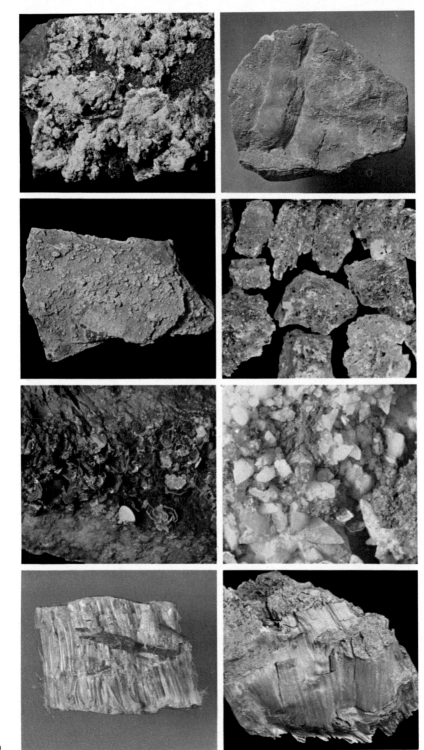

No.	Colour / Streak	Lustre / Transparency	Hardness / Specific Gravity	Cleavage / Fracture & o. / Phys. Props.	Common Form, Aggregates / Crystalline Syst.	Occurrence / Assoc. Minerals / Similar Minerals	Name & Chem. Formula / Origin of Specimen
257	dark green / light green	vitreous, pearly / transpar., transluc.	2.5 / 3.1	very good / conchoidal, brittle	thin, lamellar & needle-shaped crystals; flabelliform, rose-shaped agg. / monoclinic	decomposition product of copper ores / gypsum, malachite, azurite / —	**DEVILLINE** $CaCu_4[(OH)_3(SO_4)]_2 \cdot 3H_2O$ (Herrengrundite) / Špania Dolina, Czechoslovakia
258	emerald-green, grass-green, apple-green	pearly, vitreous / transpar., transluc.	2–2.5 / 3.4–3.6	perfect / soft to brittle	thin-tabular, lamellar micaceous crystals; scaly clusters / tetragonal	weathering product of uranium ores / partly autunite, partly torbernite / —	**CHALCOLITE** $(Cu,Ca)[UO_2 PO_4]_2 \cdot (8-10)H_2O$ (Torbernite, Copper Uranite) / Redruth, Cornwall, England
259	green, yellowish-green / yellowish-green	pearly / transpar., transluc.	2–3 / 3.5	very good / —	thin-tabular, needle-shaped crystals; bunchy, radiate agg.; crusty / tetragonal	in pegmatites, in quartz veins / uranium minerals / other uranites, cf. No. 136	**URANO-CIRCITE** $Ba[UO_2 PO_4]_2 \cdot 10H_2O$ (Uranite, Uran Mica) / Falkenstein, Vogtland, Germany
260	green-yellow, yellowish-green, sulphur-yellow / yellowish, greenish	pearly, vitreous / transluc.	2–2.5 / 3–3.2	very good / soft	thin- to thick-tabular crystals; lamellar agg.; massive / tetragonal	secondary uranium mineral, in pegmatites, in uranium veins / — / other uranium micas, cf. No. 138	**AUTUNITE** $Ca[UO_2 PO_4]_2 \cdot 10H_2O$ (Lime Uranite) / Tinh-Tuc, Tonkin, Vietnam
261	emerald-green, grass-green, apple-green, light green	pearly / transluc.	2–2.5 / 3.4–3.6	very good / brittle	thin-tabular, scaly crystals; flabelliform or rose-shaped bunches; also earthy / tetragonal	in ore veins as secondary uranium mineral, in granitic rocks in cracks & as impregnations / — / autunite, zeunerite	**TORBER-NITE** $Cu[UO_2 PO_4]_2 \cdot (8-12)H_2O$ (Copper Uranite) / Smrkovec, Czechoslovakia
262	deep verdigris green, bluish-green / greenish, white	vitreous, greasy / transpar., transluc.	2.5–3 / 6.4	good / brittle	small, columnar or needle-shaped crystals; bunchy agg. / orthorhombic	weathering product in oxidation zone of copper-lead veins / malachite, anglesite, cerussite / —	**CALEDO-NITE** $Pb_5Cu_2[(OH)_6 CO_3 (SO_4)_3]$ / Leadhills, Scotland
263	greenish-blue to blue / blue-green	vitreous / transluc.	2.5–3 / 3.5	good / brittle	short-prismatic, tabular crystals; lamellar, fibrous agg.; encrustations / orthorhombic	as weathering product in copper veins / — / —	**LANGITE** $Cu_4[(OH)_6 SO_4] \cdot H_2O$ / Špania Dolina, Czechoslovakia
264	green, blue-green / light green	dull / opaque	2–4 / 2.2–2.7	— / uneven	massive, cryptocrystalline masses, crusts; compact, botryoidal, reniform, earthy, often with agate lamination / monoclinic	as weathering product of olivine rocks, in serpentinites / opal, chrysoprase, magnesite, limonite / pimelite	**GARNIERITE** $(Ni,Mg)_6[(OH)_6Si_4O_{11}] \cdot H_2O$ (Noumeite) / Nouméa, New Caledonia

No.	Colour / Streak	Lustre / Transparency	Hardness / Specific Gravity	Cleavage / Fracture & o. / Phys. Props.	Common Form, Aggregates / Crystalline Syst.	Occurrence / Assoc. Minerals / Similar Minerals	Name & Chem. Formula / Origin of Specimen
265	green, greenish-blue, brown / pale green, greenish-blue	dull, greasy transluc., opaque	2–4 / 2–2.3	— / conchoidal, brittle to soft	cryptocrystalline, compact, reniform, botryoidal, massive, earthy, as crusts or encrustations / monoclinic	in oxidation zone of copper ore deps. / malachite, dioptase, cuprite, cerussite, limonite, azurite / aurichalcite	CHRYSO-COLLA Cu_4H_4 $[(OH)_8 Si_4O_{10}]$ / Coquimbo, Chile
266	dark green, emerald-green, blackish-green / apple-green	adamantine, vitreous transpar., transluc.	3 / 3.7	very good / brittle	columnar, needle-shaped, capillary, thick-tabular crystals; compact, massive, earthy, granular, as encrustations / orthorhombic	in oxidation zone of copper ore deps., often in volcanoes / malachite, dioptase, cuprite, haematite, limonite / malachite, olivenite	ATACAMITE $Cu(OH)Cl$. $Cu(OH)_2$ (Remolinite) / Bura-Bura, Australia
267	dark green, olive-green, yellow-brown / green-yellow, brown	vitreous, greasy, silky transluc.	3 / 4.3	poor / conchoidal, brittle	columnar, needle-shaped, tabular crystals in druses; fibrous, reniform, botryoidal, compact, earthy / orthorhombic	in oxidation zone of copper ore deps. / chalcopyrite, arsenopyrite / libethenite	OLIVENITE $Cu_2[OH As O_4]$ / St.Day, Cornwall, England
268	emerald-green / greenish	vitreous, greasy, transpar., transluc.	3.5 / 2.6	— / conchoidal, brittle	small, prismatic crystals; mostly massive, earthy, globular, stalactitic, crusts, encrustations, coatings / cubic	decomposition product of nickel ores, as coatings to chromite / brucite, clinochlore, chrysolite, serpentine / —	ZARATITE $Ni_3[(OH)_4$ $CO_3]$. $4H_2O$ (Emerald Nickel) / Texas, Pennsylvania, USA
269	dark green, brown, black / green-yellow, yellow	adamantine, vitreous, greasy transpar., transluc.	3.5–4.5 / 4–4.3	good / brittle	cube-shaped, pseudocubic, rhombohedral crystals / trigonal	in oxidation zone of iron ore deps. / limonite / —	BEUDANTITE $PbFe_3[(OH)_6$ $SO_4AsO_4]$ / Dernbach, Westerwald, Germany
270	green, yellow, brown, white-grey, yellow-red / white, yellowish, grey	adamantine, greasy transluc.	3.5–4 / 6.7–7	— / uneven, brittle	thick-tabular, columnar, barrel-shaped, ventricular, needle-shaped crystals; compact, massive, reniform, botryoidal, crusts, encrustations, coatings, also fibrous / hexagonal	in weathering zone of lead ore deps. / galena, cerussite, mimetesite, descloizite, barytes, limonite, quartz / mimetesite, vanadinite, campylite, cf. No. 325	PYRO-MORPHITE $Pb_5[Cl(PO_4)_3]$ (Green Lead Ore) / Hofsgrund, Black Forest, Germany
271	emerald-green, leek-green / green, bluish-green	vitreous transluc.	3.5 / 3.4	indistinct / uneven, brittle	short-columnar, thick-tabular, vertically striated crystals; crusts, druses / orthorhombic	in oxidation zone of copper ore deps., in mica schists / olivenite, malachite, azurite / dioptase	EUCHROITE $Cu_2[OHAsO_4]$ $3H_2O$ / Lubětová, Czechoslovakia
272	white, yellow, green, brown, bluish / white	vitreous, silky transluc.	3.5–4 / 2.3–2.4	good / brittle	fine-fibrous, radiate bunches; hemispheric agg. / orthorhombic	in cracks in sandstone, silica schists, quartzite / — / prehnite, cf. No. 68	WAVELLITE $Al_3[(OH)_3$ $(PO_4)_2$. $5H_2O$ / Montgomery Co., Arkansas, USA

No.	Colour / Streak	Lustre / Transparency	Hardness / Specific Gravity	Cleavage Fracture & o. / Phys. Props.	Common Form, Aggregates / Crystalline Syst.	Occurrence / Assoc. Minerals / Similar Minerals	Name & Chem. Formula / Origin of Specimen
273	emerald-green, dark green, greenish-black / light green	vitreous, silky, dull / transluc., opaque	$\dfrac{4}{4.0}$	good / fibrous separation, brittle	fibrous, needle-shaped, capillary crystals; bunchy agg.; compact, reniform; botryoidal, nodular, stalactitic, subvitreous, agate-like striated, massive, earthy, disseminated, impregnations, encrustations, coatings / monoclinic	in weathering zone of copper deps.; impregnation in sandstone / chalcopyrite, bornite, tetrahedrite, chalcosine, cuprite, copper, azurite, haematite, limonite / chrysocolla, pseudomalachite, atacamite	**MALACHITE** $Cu_2[(OH)_2CO_3]$ / Moldova, Banat, Romania
274							**MALACHITE** $Cu_2[(OH)_2CO_3]$ / Katanga, Zaïre
275	dark green, greenish-black / light green	greasy, vitreous / transluc.	$\dfrac{4}{3.8}$	— / conchoidal, brittle	small, short-columnar, octahedral crystals; reniform, globular, radiate, fibrous, bunchy, massive / orthorhombic	in oxidized zone of copper ores / malachite, cuprite, limonite, quartz / malachite, pseudomalachite, olivenite, atacamite	**LIBETHE-NITE** $Cu[OH\ PO_4]$ / Lubětová, Czechoslovakia
276	green, bluish-green, colourless / white	greasy / transluc.	$\dfrac{4-5}{2.2-2.5}$	— / conchoidal	small, short-columnar, needle-shaped crystals; globular agg.; massive, crusty, reniform / orthorhombic	in fissures and seams of silica schists, quartzite / — / wavellite	**VARISCITE** $Al[PO_4] . 2H_2O$ / Montgomery Co., Arkansas, USA
277	emerald-green, dark green, bluish-green	vitreous, greasy / transluc., transpar.	$\dfrac{4-5}{3.8-4.4}$	indistinct conchoidal, brittle	short-columnar, fibrous, grown-up crystals; radiate-fibrous, reniform-botryoidal agg. / monoclinic	in oxidized zone of copper deps., in fissures in limestones, dolomites, sandstone / malachite, azurite, limonite, chalcedony	**PSEUDO-MALACHITE** $Cu_5[(OH)_2\ PO_4]_2$ / Lubětová, Czechoslovakia
278	dark green, emerald-green, apple-green	dull / transluc., opaque	$\dfrac{4.5-5}{4-4.1}$	— / conchoidal, brittle	fine needle-shaped spherulites; reniform, shelled agg. / monoclinic	in oxidized zone of copper deps. / olivinite / malachite	**CORN-WALLITE** $Cu_5[(OH)_2\ As\ O_4]_2$ (Erinite) / Carrarach Mine, Cornwall, England
279	dark green, brownish, brown-green / light green, brownish	pearly, greasy / transluc., opaque	$\dfrac{4-4.5}{3-3.2}$	very good / brittle	tabular, columnar, grown-up crystals; granular agg.; compact / hexagonal	metasomatic structure in iron & manganese deps. / — / —	**PYRO-SMALITE** $(Mn,Fe)_8[(OH,Cl)_{10}Si_6O_{15}]$ / Broddebo, Kalmar, Sweden
280	white, yellow, red, orange-red, green, grey, bluish / white	vitreous, greasy / transluc.	$\dfrac{5}{4.3-4.5}$	very good / brittle	small rhombohedral crystals; often rounded, mostly compact, botryoidal, reniform, conical, finely granular, massive, earthy, shell-like, striated sinter crusts, chalcedonic / trigonal	metasomatic in limestone and dolomite rocks; as isolated masses in fissures and cavities, weathering product of sphalerite / galena, hydrozincite, malachite & o. / hemimorphite, prehnite, chalcedony, cf. No. 411	**SMITH-SONITE** $ZnCO_3$ (Calamine, Zinc Spar) / Tsumeb, South-West Africa

No. 273 – 280   **Table 3**

No.	Colour Streak	Lustre Transparency	Hardness Specific Gravity	Cleavage Fracture & o. Phys. Props.	Common Form, Aggregates Crystalline Syst.	Occurrence Assoc. Minerals Similar Minerals	Name & Chem. Formula Origin of Specimen
281	colourless, white, yellow, red, greenish, violet / white	vitreous, greasy / transpar., transluc.	5 / 3.1	clear / conchoidal, brittle	hexagonal, short- to long-columnar, tabular crystals; compact, reniform, nodular, massive, granular / hexagonal	as constituent in magmatic rocks, in fissures in crystalline schists, in pegmatites / — / beryl, nepheline, cf. No. 82, 158, 214	APATITE $Ca_5[F (PO_4)_3]$ Slavkov, Czechoslovakia
282	light green, blue-green / white	greasy to dull / opaque	5 / 2.8	— / brittle	compact, concentric spherules, reniform, stalactitic, nodular, disseminated, massive / tetragonal	crusts in cavities in decomposed variscite, in silica schists / — / —	WARDITE $NaAl_3 [(OH)_4(PO_4)_2]$ $2H_2O$ Lewiston, Utah USA
283	colourless, greenish, yellowish, grey, brown, bluish-red / white	vitreous, greasy / transpar., transluc.	5.5 / 4—4.2	good / conchoidal, splintery	small, needle-shaped crystals; compact, coarsely to finely granular / trigonal	in manganese & zinc deps. / smithsonite, hemimorphite, zincite, franklinite, rhodonite, limonite, calcite / olivine, epidote	WILLEMITE $Zn_2(SiO)_4$ Franklin, New Jersey, USA
284	emerald-green, dark green / bluish-green	vitreous / transpar., transluc.	5 / 3.3	very good / conchoidal, brittle	short-columnar, grown-up crystals; in druses; massive agg. / trigonal	in oxidized zone of copper lodes, in fissures in limestone, dolomite, sandstone / chrysocolla, limonite / malachite, atacamite	DIOPTASE $Cu_6[Si_6O_{18}] \cdot 6H_2O$ Rénéville, Zaïre
285	emerald-green / greenish	vitreous / transpar., transluc.	5—6 / 3.2—3.3	good / conchoidal, brittle	compact, only granular / monoclinic	in crystalline schists & eclogites / garnet, zoisite, cyanite, epidote / fassaite	OMPHACITE (Ca,Na) (Mg,Fe,Al) $[Si_2O_6]$ (Pyroxene Family) Ötztal, Tyrol, Austria
286	colourless, yellowish, green, greenish-black / white	vitreous, greasy / transpar., transluc.	5—6 / 3.2—3.5	clear, good / brittle	short-columnar, needle-shaped, grown-up crystals; compact, granular, tubular, shelled, massive / monoclinic	in magnetite lodes, in fissures in metamorphic rocks / chlorite, hessonite, magnetite, apatite, biotite / clinochlore, augite	DIOPSIDE $CaMg[Si_2O_6]$ (Pyroxene Family) Rothenkopf, Zillertal, Tyrol, Austria
287	green, grey-green / white	dull / transluc.	5.5 / 2.9—3	— / very massive & tenacious	cryptocrystalline, slaty, only compact, massive / monoclinic	compact actinolite or also anthophyllite, in loose blocks & pebbles of serpentine gabbro / — / jadeite	NEPHRITE $(Mg,Fe)_7[OH Si_4O_{11}]_2Na_2Ca_2 (Mg,Fe)_{10} [(OH)_2O_2Si_{16}O_4$ Kawa-Kawa, New Zealand
288	grey-green, grey-white, light green, white, brown / white	pearly, silky / transluc.	5—6 / 3.1	good / fibrous separation, rather brittle	fine, needle-shaped, capillary (up to 15 cm long) crystals; parallelly fibrous, felt-like agg. / monoclinic	asbestos variety of cummingtonite / — / chrysotile-asbestos	AMOSITE $(Mg,Fe)_7 [OH Si_4 O_{11}]_2$ Lydenburg, Transvaal, South Africa

No.	Colour / Streak	Lustre Transparency	Hardness / Specific Gravity	Cleavage Fracture & o. Phys. Props.	Common Form, Aggregates / Crystalline Syst.	Occurrence Assoc. Minerals / Similar Minerals	Name & Chem. Formula / Origin of Specimen
289	green, reddish / white	vitreous, metallic glint transluc.	5.5−6 / 2.6	— uneven	compact, finely granular to massive, slaty / trigonal	— / spangles of mica or haematite / —	Quartz variety AVENTURINE SiO_2 / Brazil
290	green, grey / white	vitreous, silky transpar., transluc.	5.5−6 / 2.9−3.2	good brittle	long-acicular, capillary crystals; fine-fibrous, felt-like asbestiform agg. / monoclinic	in metamorphic limestones & dolomites, in alpine cracks / — / —	Amphibole variety BYSSOLITE $Ca_2Mg_5(OH,F)[Si_4O_{11}]_2$ / Knappenwand, Salzburg, Austria
291	colourless, white, yellow, greenish, grey-green, blue-green, brown, red / white	vitreous, greasy transpar., transluc.	5.5−6 / 2.6	indistinct conchoidal to uneven	short-columnar to thick-tabular crystals; mostly massive, compact, granular / hexagonal	rock constituent in volcanic ejecta / — / apatite, quartz, felspars	NEPHELINE $KNa_3[AlSiO_4]_4$ / Vesuvius, Italy
292	grey, yellow, greenish, light rose, red / white	vitreous transluc., opaque	6.5 / 3.2	very good uneven	long-columnar, tubular, thick-tabular crystals; often twinned / monoclinic	in metamorphic rocks & crystalline schists / — / zoisite, tremolite	CLINOZOISITE $Ca_2Al_3[O\,OH\,SiO_4\,Si_2O_7]$ / Ampandandrava, Madagascar
293	grey-white, greenish, brownish, pink, red / white	vitreous, pearly opaque	6−6.5 / 3.2−3.3	very good uneven	columnar to acicular crystals; compact, tubular, radial agg. / orthorhombic	in crystalline schists & metamorphic rocks / amphibole, garnet, vesuvianite, epidote, quartz & o. / tremolite	ZOISITE $Ca_2Al_3[O\,OH\,SiO_4\,Si_2O_7]$ / Zermatt, Switzerland
294	green, greenish-white, grey / white	vitreous, pearly transpar., transluc.	6−6.5 / 2.8−3	clear uneven	tabular, short-columnar crystals; globular, reniform, shelled, flabelliform agg.; compact, granular, crusts / orthorhombic	in fissures & vesicles of basic igneous rocks & crystalline schists / zeolite, calcite, axinite / hemimorphite, aragonite, staffelite	PREHNITE $Ca_2Al[(OH)_2 AlSi_3O_{10}]$ / West Patterson, New Jersey, USA
295	dark green, yellow-green, yellow, brown, red, grey, black / white, grey	vitreous transluc.	6−7 / 3.3−3.5	very good conchoidal, splintery	columnar, acicular, transversely striated crystals; compact, platy, radial, bunchy agg.; granular to massive / monoclinic	in alpine cracks, in crystalline schists, in metamorphic limestone, in fissures in basalt	EPIDOTE $Ca_2(Fe,Al)Al_2[O\,OH\,SiO_4 Si_2O_7]$ / Knappenwand, Untersulzbachtal, Salzburg, Austria
296	pistachio-green, yellow-green / white	vitreous transluc.	6−7 / 3.3−3.5	very good conchoidal, splintery	columnar, acicular crystals; tubular, radial agg.; granular, compact, massive / monoclinic	garnet, vesuvianite, copper / tourmaline, actinolite, vesuvianite	Epidote variety PISTACITE $Ca_2(Fe,Al)Al_2[O\,OH\,SiO_4 Si_2O_7]$ / Erbendorf, Fichtelgebirge, Germany

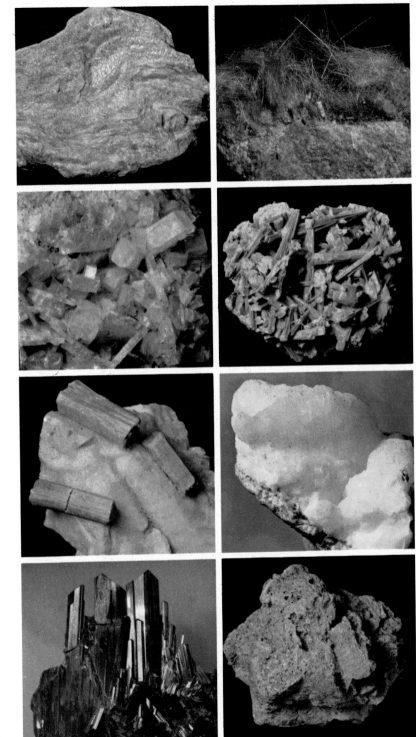

No.	Colour Streak	Lustre Transparency	Hardness Specific Gravity	Cleavage Fracture & o. Phys. Props.	Common Form, Aggregates Crystalline Syst.	Occurrence Assoc. Minerals Similar Minerals	Name & Chem. Formula Origin of Specimen
297	light green, apple-green white	vitreous, greasy transluc.	7 / 2.6	— con-choidal	massive, reniform, shelled, dissemi-nated, crypto-crystalline / trigonal	in nickel ore deps. — prehnite	Quartz variety **CHRYSO-PRASE** SiO_2 (coloured green due to presence of nickel com-pounds) Szklary, Poland
298	dark green, dirty green, leek-green whitish	vitreous, greasy opaque, transluc.	7 / 2.5 — 2.6	— con-choidal	irregular nodules, veins, disseminated, cryptocrystalline; massive / trigonal	in igneous rocks, in serpentines quartz, chalcedony prase	Quartz variety **PLASMA** SiO_2 Hrubšice, Czechoslovakia
299	yellow-green, olive-green, greenish-black, red-brown white	vitreous, greasy transpar., transluc.	6.5 — 7 / 3.3	good con-choidal	thick-tabular, short-columnar crystals; granular to massive, compact, granular bunches, xenoliths orthorhombic	constituent in basic igneous rocks in meteorites, in magnetite lodes — chrysoberyl, cf. No. 354	**OLIVINE** $(Mg,Fe)_2[SiO_4]$ (Chrysolite, Peridot) St.John's Island, Red Sea
300	green, yellow-green white	vitreous, ada-mantine transpar., transluc.	7 / 3.3 — 4.1	— con-choidal, splintery	dodecahedral, mostly grown-up crystals; massive cubic	in fissures in crys-talline schists, in ore deps., in gold placers —	Andradite variety **DEMANTOID** $Ca_3Fe_2[SiO_4]_3$ (Garnet Family) Frascati, Albanian Mountains, Italy
301	greenish-white, green, grey, yellowish white	vitreous, greasy transluc.	6.5 — 7 / 3.2 — 3.3	— very tenacious, uneven	only compact, massive, granular, fibrous agg.; nodules, pebbles, cryptocrystalline monoclinic	in crystalline schists — nephrite	**JADEITE** $NaAl[Si_2O_6]$ Mogok, Upper Burma
302	green white	vitreous transpar., transluc.	7 / 3 — 3.2	— uneven, splintery	columnar to long-columnar crystals; tubular to fibrous agg.; granular, massive trigonal	in pegmatites, granites & metamorphic rocks — epidote, vesuvianite	Tourmaline variety **VERDELITE** $NaFe_3Al_6$ $[(OH)_4 (BO_3)_3$ $Si_6O_{18}]$ Dravograd, Yugoslavia
303	emerald-green, dark green whitish	vitreous, greasy transluc., transpar.	7.5 — 8 / 3.4	— con-choidal, splintery	dodecahedral crys-tals; granular to massive, as crusts cubic	in serpentines rich in chromite, in ore deps. — dioptase	**UVAROVITE** $Ca_3Cr_2[SiO_4]_3$ (Garnet Family) Biser, Central Urals, USSR
304	yellow, light green, emerald-green white	vitreous, greasy transpar., transluc.	8.5 / 3.7	clear con-choidal	tabular, short-columnar, tubular crystals; often twinned, granular orthorhombic	in granitic pegma-tites, granites, syenites, gneiss & mica schists — olivine, beryl, apatite, tourmaline, corundum	**CHRYSO-BERYL** Al_2BeO_4 Haddam, Connecticut, USA

No.	Colour / Streak	Lustre / Transparency	Hardness / Specific Gravity	Cleavage / Fracture & o. / Phys. Props.	Common Form, Aggregates / Crystalline Syst.	Occurrence / Assoc. Minerals / Similar Minerals	Name & Chem. Formula / Origin of Specimen
305	greenish, yellow-green, blue-green, yellow, red / white	vitreous, dull / transpar., transluc.	7.5 – 8 / 2.6	almost indistinct / conchoidal to uneven	hexagonal, columnar, tabular to acicular crystals; tubular, fibrous agg.; compact / hexagonal	in granitic pegmatites, granites, crystalline schists, in tin deps. / tourmaline, topaz / apatite, tourmaline	**BERYL** $Al_2Be_3[Si_6O_{18}]$ / Meclov, Czechoslovakia
306	dark green, bluish-green / white	vitreous / transpar., transluc.	7.5 – 8 / 2.6	almost indistinct / conchoidal to uneven	hexagonal, columnar crystals / hexagonal	crystals embedded in mica schists, limestone / — / cf. No. 516	Beryl variety **EMERALD** $Al_2Be_3[Si_6O_{18}]$ / Muzo, Colombia

No.	Colour / Streak	Lustre / Transparency	Hardness / Specific Gravity	Cleavage / Fracture & o. / Phys. Props.	Common Form, Aggregates / Crystalline Syst.	Occurrence / Assoc. Minerals / Similar Minerals	Name & Chem. Formula / Origin of Specimen
307	brownish, grey, black / brown, grey-black	metallic, dull / opaque	2.5 / 6.2	— / brittle to soft	pseudocubic, small cube-shaped crystals; compact, reniform, botryoidal, mammillary, massive / orthorhombic	in silver veins / pyrargyrite, galena, sphalerite, siderite / germanite	**ARGYRODITE** $4Ag_2S . GeS_2$ / Freiberg, Ore Mountains, Germany
308	gold-brown, tombac-brown, dark brown / grey-black	metallic / opaque	4 / 4.6	clear / brittle, uneven	thick-tabular to short-columnar crystals; compact, granular, massive, in masses / hexagonal	in basic igneous rocks, in ore veins / — / cf. No. 128	**PYRRHOTITE** FeS (Magnetic Pyrites) / Dell Norte, California, USA

No.	Colour / Streak	Lustre / Transparency	Hardness / Specific Gravity	Cleavage / Fracture & o. / Phys. Props.	Common Form, Aggregates / Crystalline Syst.	Occurrence / Assoc. Minerals / Similar Minerals	Name & Chem. Formula / Origin of Specimen
309	colourless, white, grey, yellow, brown / white	adamantine, greasy / transpar., transluc.	2.5 – 3 / 6 – 6.3	perfect / conchoidal, brittle to soft	short-columnar, tabular, mostly grown-up crystals / tetragonal	in weathering zone of lead veins / — / anglesite, cerussite	**PHOSGENITE** $Pb_2[Cl_2O_3]$ (Horn Lead) / Monte-Poni, Sardinia
310	brown, white, grey-white, yellow-brown, black / whitish	greasy to dull / opaque	2 – 5 / 3 – 3.2	— / brittle	radial, fibrous, globular agg.; compact, reniform, nodular, earthy, granular, loose, cryptocrystalline / hexagonal	microcrystalline apatite from sedimentary deps. / — / —	Apatite variety **PHOSPHORITE** $[3CaPo_4]_x$ $[(2Ca F_2 CO_3)]_y$ $2H_2O$ / Děražnja, Podolien, Ukraine, USSR
311	light brown, silver-grey, dark grey, yellowish, greenish / white	pearly, metallizing / transpar., transluc.	2.5 – 3 / 2.9 – 3.1	perfect / soft to brittle, elastic, flexible, foliaceous	tabular, thin-lamellar, pseudo-hexagonal crystals; loose groups; scaly, lamellar, tubular agg. / monoclinic	in tin ore deps., in granites & gneiss / cassiterite, wolframite, scheelite, fluorite, pycnite, quartz / lepidolite	**ZINNWALDITE** $KLiFeAl$ $[(F,OH)_2$ $AlSi_3O_{10}]$ (Lithionite) / Cínovec, Czechoslovakia
312	colourless, grey, brown, green-yellow, silver-white / white	pearly, vitreous, silky / transpar., transluc.	2 – 2.5 / 2.7 – 2.8	perfect / elastic, flexible, foliaceous	thin- to thick-tabular, short-columnar, pseudo-hexagonal crystals; thin-lamellar, compact, scaly, globular, massive / monoclinic	original constituent of several kinds of rocks, except basic rocks, in tin deps. / — / other micas	**MUSCOVITE** $KAl_2[(OH,F)_2$ $AlSi_3O_{10}]$ (Mica Family) / Middletown, Connecticut, USA

No.	Colour Streak	Lustre Transparency	Hardness Specific Gravity	Cleavage Fracture & o. Phys. Props.	Common Form, Aggregates Crystalline Syst.	Occurrence Assoc. Minerals Similar Minerals	Name & Chem. Formula Origin of Specimen
313	gold-brown, yellow-green, grey, blackish _____ greenish	vitreous, pearly, metallic _____ transpar., transluc.	$1-2$ $\overline{2.1-2.7}$	perfect soft, inflates strongly when heated	pseudohexagonal thin & thick laminae; scaly, massive agg. _____ monoclinic	weathering product of micas _____ phlogopite, chlorite, magnesite _____ biotite, phlogopite	**VERMICULITE** Mg_3FeAl $[(OH)_2$ $AlSi_2O_{10}Mg$ $(H_2O)_4]$ _____ Westchester, Pennsylvania, USA
314	dark brown, deep green, black _____ dark olive-green	vitreous, metallic _____ opaque	$\overline{3.5}$ $\overline{3.4}$	very good elastic	triangular, mostly hexagonal, thin, foliaceous crystals; globular clusters; tubular, radiate-fibrous agg.; reniform _____ monoclinic	in ore veins _____ pyrite, siderite, limonite, sphalerite, quartz _____	**STILPNOMELAN** (K,H_2O) $(Fe,Mg,Al)_3$ $[(OH)_2Si_4O_{10}]$ $(H_2O)_2$ _____ Chabičov, Czechoslovakia

No.	Colour Streak	Lustre Transparency	Hardness Specific Gravity	Cleavage Fracture & o. Phys. Props.	Common Form, Aggregates Crystalline Syst.	Occurrence Assoc. Minerals Similar Minerals	Name & Chem. Formula Origin of Specimen
315	colourless, yellow, red, brown, green, grey, black _____ white, yellow, brown	adamantine, greasy _____ transpar. to opaque	$3.5-4$ $\overline{3.9-4.2}$	very good, rhombo-dodecahedral _____ brittle	tetrahedral, dode-cahedral, cube-like crystals; coarse to fine-granular, massive, compact _____ cubic	in ore veins _____ galena, chalcopyrite & o. _____ tetrahedrite, cassiterite, garnet, cf. No. 142, 191, 445	**SPHALERIT** ZnS (Zinc Blende) _____ Cumberland, England
316	light brown, dark brown _____ brownish, yellow	adamantine, resinous _____ transluc., opaque	$3-4$ $\overline{4}$	perfect _____ brittle	columnar, acicular crystals; tubular to fibrous agg.; crusty, shelled, striated, massive _____ hexagonal	in lead-zinc ore deps. _____ sphalerite _____	**WURTZITE** ZnS (Radial Blende) _____ Příbram, Czechoslovakia
317	brown, yellowish, brown-green, grey, brown-black _____ white, yellowish	vitreous, pearly, dull _____ transluc.	$3.5-4$ $\overline{3.7-3.9}$	perfect, rhombo-hedral _____ brittle	rhombohedral, lenticular, often curved, distorted crystals; saddle-shaped agg.; coarsely to finely granular, spathic, massive _____ trigonal	in metasomatic ore deps., in ore veins, sedimentary _____ clay _____ calcite, dolomite, ankerite, magnesite	**SIDERITE** $FeCO_3$ (Chalybite, Spathose Iron) _____ Neudorf near Quedlinburg, Harz, Germany
318	light brown, yellowish-white _____ white	dull _____ opaque	$3.5-4$ $\overline{2.8}$	— _____ brittle, uneven	globular, reniform, botryoidal, shelled agg.; crypto-crystalline _____ trigonal	in sedimentary rocks & gypsum _____ — _____ —	**MIEMITE** $CaMg[CO_3]_2$ (Dolomite) _____ Durham, England
319	brown, yellow-brown, whitish-grey _____ white	dull _____ opaque	3 $\overline{2.6-2.8}$	— _____ uneven	massive, solid masses, banded, cryptocrystalline _____ trigonal	massive limestones with angular patterns, unevenly coloured by iron hydroxide _____ — _____ —	Calcite variety **RUIN MARBLE** $CaCO_3$ _____ Florence, Italy
320	yellow, white, brown, red-brown _____ white, yellowish	dull, silky _____ opaque	$3.5-4$ $\overline{2.9-3}$	indistinct, fibrous, jointing _____ brittle	radiating & parallely fibrous agg.; stalactitic, concentrically shelled, coatings, crusts, sinter formation _____ orthorhombic	aragonite sinter as deposition of hot springs _____ — _____ cf. No. 485	Aragonite variety **SPRUDELSTEIN** $CaCO_3$ _____ Carlsbad, Czechoslovakia

No.	Colour / Streak	Lustre / Transparency	Hardness / Specific Gravity	Cleavage Fracture & o. / Phys. Props.	Common Form, Aggregates / Crystalline Syst.	Occurrence Assoc. Minerals / Similar Minerals	Name & Chem. Formula / Origin of Specimen
321	white, yellowish, brown, grey, black / white	adamantine, greasy / transpar., transluc.	3—3.5 / 6.4—6.6	indistinct / conchoidal, brittle	columnar, tabular, acicular crystals; often twins, cruciform or radiate arrangements, compact, reniform, massive, tubular, / orthorhombic	in weathering zone of lead veins / galena, anglesite, pyromorphite / anglesite, phosgenite, coelestine, barytes, scheelite, cf. No. 401	CERUSSITE $PbCO_3$ (White Lead Ore) / Příbram, Czechoslovakia
322	yellow, brown, blackish-brown / straw-yellow	vitreous, dull / opaque	3—4 / 2.9—3.2	clear / brittle	very small, rhombohedral, tabular crystals; mostly compact, powdery, reniform, granular, crusty, fibrous / trigonal	in weathering zone of ore veins, in clays / limonite, haematite, alunite, quartz / different ochres	JAROSITE $KFe_3[(OH)_6(SO_4)_2]$ / Laurion, Greece
323	yellowish, orange, red, grey, brown, green / white	adamantine, greasy / transluc., opaque	3 / 7.9—8.3	indistinct / brittle to soft	pointed-pyramidal, short-columnar crystals; bundle- & fusiform, grown-up granules / tetragonal	in tin ore deps. / scheelite / scheelite, cf. No. 171	STOLZITE $[PbWO_4]$ / Cínovec, Czechoslovakia
324	brown, brown-red, brownish-green, black / yellow, yellow-brown	adamantine, greasy / transluc., opaque	3.5 / 5.9—6.2	— / conchoidal, brittle	columnar, pyramidal crystals; mostly crusty forms, botryoidal, mammiform, radiate, coatings / orthorhombic	in lead-zinc ore deps. / pyromorphite, vanadinite, limonite, cerussite / —	MOTTRAMITE $Pb(Cu,Zn)[OH VO_4]$ / Friesenberg, South-West Africa
325	brown, green, yellow, yellow-red, / white, yellowish, grey	adamantine, greasy / transluc.	3.5—4 / 6.7—7	— / uneven, brittle, conchoidal	columnar, barrel-shaped, acicular, ventricular crystals; reniform, botryoidal, compact, massive, crusts, coatings / hexagonal	in weathering zone of lead deps. / — / mimetesite, campylite, vanadinite, cf. No. 270	PYRO-MORPHITE $Pb_5[Cl(PO_4)_3]$ / Oloví, Czechoslovakia
326	yellow, brown, reddish-brown, orange, ruby-red / yellow, brownish	adamantine, greasy / transluc., opaque	3 / 6.8—7.1	— / conchoidal, brittle	short-columnar, barrel-shaped crystals; reniform, botryoidal, compact, fibrous, thin-tubular agg. / hexagonal	in lead deps. / wulfenite, galena, pyromorphite, descloizite & o. / pyromorphite, mimetesite, campylite	VANADINITE $Pb_5[Cl(VO_4)_3]$ / Djebel Mahser Morocco
327	brown, yellow, orange, green / yellow	adamantine, greasy / transluc.	3.5 / 7.2	good / brittle	barrel-shaped, ventricular, thick-tabular crystals; hemispheric, bundle-shaped agg. / hexagonal	in lead deps. / — / —	Mimetesite variety CAMPYLITE $Pb_5[Cl(As,P)O_4]_3$ (Phosphoreous Mimetesite) / Příbram, Czechoslovakia
328	brown, brass-brown, gold-yellow / yellow-brown	adamantine, pearly, metallizing / transpar., transluc.	3—4 / 3.3—3.4	perfect / brittle	tabular, foliaceous crystals; scaly, rose-shaped, stellate agg. / triclinic	embedded in nepheline syenites / amphibole, mica, zircon / —	ASTRO-PHYLLITE $(K,Na)_3(Fe,Mn)_7(Ti,Zr)_2[Si_8(O,OH)_{31}]$ / Los Insel, West Africa

No.	Colour / Streak	Lustre / Transparency	Hardness / Specific Gravity	Cleavage / Fracture & o. / Phys. Props.	Common Form, Aggregates / Crystalline Syst.	Occurrence / Assoc. Minerals / Similar Minerals	Name & Chem. Formula / Origin of Specimen
329	brown-green, black, earthy on surface, yellow / yellowish, white	greasy, waxy, dull on surface / opaque	$\frac{4}{3.6-3.7}$	— / conchoidal, brittle	octahedral, dodecahedral, large crystals; compact, granular, massive / cubic	in granite-pegmatites / euxenite, zircon, biotite & o. / —	**BETAFITE** $(Ca,Th,U,Ce)(Nb,Ta,Ti)_3O_9 \cdot nH_2O$ / Betafo, Madagascar
330	dark brown, brownish-yellow, grey / whitish	dull / opaque	$\frac{4}{3.7-3.9}$	— / brittle	globular, reniform, nodular, concretions; massive, cryptocrystalline / trigonal	argillaceous siderite, in sedimentary rocks, in coal seams / — / —	Siderite variety **PELO-SIDERITE** $FeCO_3$ / Cumberland, England
331	yellowish, brown, grey / white	vitreous, greasy / transluc.	$\frac{4-4.5}{2.9-3.1}$	very good / brittle	rhombohedral, lentiform crystals; spathic, granular / trigonal	embedded in chlorite schists & talc schists / — / dolomite, ankerite, siderite	**BREUN-NERITE** $(Mg,Fe)CO_3$ (Mesitine Spar, Mesitite) / Greiner Wald, Upper Austria
332	brown-yellow, red-brown, orange, black / brown, orange-yellow	vitreous, greasy / transluc., opaque	$\frac{4.5-5}{4.4-4.8}$	poor, indistinct / conchoidal, brittle	short-columnar, pyramidal, embedded crystals; massive, compact, lamellar, disseminated / tetragonal	in pegmatites / zircon, aegirine, felspars / zircon, xenotime	**THORITE** $Th(SiO_4)$ / Brevik on Langesundsfjord, Norway
333	brown, bronze-brown, grey-green / white, grey	vitreous, pearly, metallic / opaque, transluc.	$\frac{5-6}{3.2-3.5}$	good / striated cleavage planes	tabular, columnar, lamellar crystals; compact, coarsely granular, spathic, bundle-shaped, radial / orthorhombic	in basic rocks / enstatite, olivine, serpentine, chromite, magnetite / enstatite, hypersthene, diallage	**BRONZITE** $(Mg,Fe)_2[Si_2O_6]$ / Gefrees, Fichtelgebirge, Germany
334	brown, blackish-brown / brown, yellowish-brown	greasy / transluc., opaque	$\frac{5-5.5}{4.3-4.5}$	poor / conchoidal, brittle	octahedral, cube-shaped crystals; granular, disseminated / cubic	in granite-pegmatites, in carbonates / zircon, nepheline / —	**PYROCHLORE** $(Ca,Na,Y,Ce,Th,U,Ti)_2 Nb_2O_6(F,OH,O)$ / Frederiksvärn near Larvik, Norway
335	brown, ochre-yellow / dark yellow, brown	dull, velvety / opaque	$\frac{5-5.5}{3.3-4.3}$	good / fibrous, brittle	needle-shaped, capillary crystals; fine-fibrous agg.; globular, reniform, mammillary structure / orthorhombic	in oxidation zone of ore veins / — / —	**SAMT-BLENDE** $\alpha\text{-}FeOOH$ (Goethite, Lepidocrocite) / Příbram, Czechoslovakia
336	yellow, yellow-brown, rusty-brown / yellow	silky, dull / opaque	$\frac{5-5.5}{3.7}$ also softer	— / earthy, fibrous	radial, stalactitic, compact, earthy, cryptocrystalline / orthorhombic	in weathering zone of iron deps. / — / different ochres	**XANTHO-SIDERITE** $FeOOH$ / Lostwithiel, Cornwall, England

No.	Colour / Streak	Lustre / Transparency	Hardness / Specific Gravity	Cleavage / Fracture & o. / Phys. Props.	Common Form, Aggregates / Crystalline Syst.	Occurrence / Assoc. Minerals / Similar Minerals	Name & Chem. Formula / Origin of Specimen
337	brown, red, yellow, orange grey-white	ada-mantine, greasy transluc., opaque	5−5.5 / 4.8−5.5	perfect con-choidal, brittle	thick-tabular, short-columnar, embedded & grown-up crystals; rolled grains / monoclinic	in pegmatites, granites, syenites, gneiss, in alpine cracks, loose in sands, in placers / zircon, rutile, ilmenite / thorite, orthite	**MONAZITE** Ce[PO$_4$] / Ambatofotsikely Madagascar
338	greenish-blue, brownish-green, reddish-brown, blue-green white	vitreous, greasy transluc., dull	5 / 3.1−3.2	less distinct con-choidal, brittle	columnar, hexa-gonal crystals; granular, compact / hexagonal	in lithium-pegma-tites / — / cf. No. 158	Apatite variety **MOROXITE** Ca$_5$[F(PO$_4$)$_3$] / Renfrew, Quebec, Canada
339	yellow, brown, reddish, greenish, black white	vitreous, greasy transpar., transluc.	5−5.5 / 3.4−3.6	indistinct con-choidal, brittle	columnar, wedge-shaped, tabular, needle-shaped crystals / monoclinic	in alpine cracks, in crystalline schists / axinite, cf. No. 162, 463	**TITANITE** CaTi(O SiO$_4$) / Teufelsmühle, Habachtal, Salzburg, Austria
340	brown, black-grey, black white, grey	vitreous, ada-mantine transluc., opaque	6.5 / 3.6	clear brittle	long-columnar crystals; radial, rose-shaped agg. / monoclinic	in granite-pegma-tites / monazite, zircon, beryl, ilmenite, magnetite	**THORT-VEITITE** Sc$_2$[Si$_2$O$_7$] / Befanamo, Madagascar
341	brown, brownish-black, yellowish crust yellow-brown, grey-brown	greasy, sub-metallic transluc., opaque	5.5−6.5 / 4.6−5.9	— con-choidal, brittle	tubular, prismatic, tabular crystals; flabelliform agg.; granular / orthorhombic	in granite-pegma-tites / zircon, mona-zite, beryl, ilmenite, magnetite / —	**EUXENITE** (Y,Ce,U,Pb,Ca) (Nb,Ta,Ti)$_2$ (O,OH)$_6$ (Polycrase) / Ampangabe, Madagascar
342	brownish-black, grey-black blackish, black-green	vitreous, greasy, sub-metallic opaque	5.5−6 / 4.1	clear con-choidal, brittle	columnar, longi-tudinally striated, needle-shaped, capillary crystals; tubular, radial, fibrous agg.; granu-lar, compact / orthorhombic	in contact-meta-morphic rocks, in lavas / augite, amphibole / tourmaline, actino-lite	**ILVAITE** CaFe$_2$. CaFe$_3$ [OH O Si$_2$O$_7$] (Lievrite, Yenite) / Rio Marina, Elba, Italy
343	brown, dark grey, blackish-green, black white, grey	vitreous opaque	5−6 / 3.5−3.6	clear brittle	tubular, needle-shaped crystals; spathic, granular, compact, massive / monoclinic	in metamorphic & metasomatic rocks / magnetite, pyrite & o. / —	**HEDEN-BERGITE** CaFe[Si$_2$O$_6$] / Nordmark, Värmland, Sweden
344	brownish-green, green, greenish-black white, grey	vitreous transpar., transluc.	5−6 / 3.2−3.3	good brittle	short-, long- & thick-columnar to acicular crystals; compact, dissemi-nated, granular to massive / monoclinic	in cracks of meta-morphic rocks, in volcanic ejecta / garnet, spinel, vesuvianite / diopside, amphibole	**FASSAITE** Ca(Mg,Fe,Al) [(Si,Al)$_2$O$_6$] / Brosso near Traversella, Piedmont, Italy

No.	Colour / Streak	Lustre / Transparency	Hardness / Specific Gravity	Cleavage / Fracture & o. / Phys. Props.	Common Form, Aggregates / Crystalline Syst.	Occurrence / Assoc. Minerals / Similar Minerals	Name & Chem. Formula / Origin of Specimen
345	brown, grey / white	vitreous, pearly, metallic, iridescent / transluc.	5 — 6 / 3.1 — 3.2	good / soft, elastic	needle-shaped, tabular, fibrous crystals; tubular, fibrous, lamellar agg.; compact, nodules / orthorhombic	orthorhombic amphibole, in metamorphic rocks, in crystalline schists, as radiating-columnar shells, so-called 'Glimmerskugeln' (mica globules) / — / cf. No. 408	**ANTHOPHYLLITE** $(Mg,Fe)_7$ $[OH\ Si_4O_{11}]_2$ / Heřmanov, Czechoslovakia
346	brownish, reddish, grey, white, yellowish, colourless / white	vitreous, pearly / transluc., transpar.	6 — 6.5 / 2.5	very good / brittle, uneven	short-columnar, thick platy crystals; spathic, granular, massive / monoclinic	main constituent of numerous rocks / — / cf. No. 205	**ORTHOCLASE** $K(AlSi_3O_8)$ (Felspar Family) / Strigom, Poland
347	brown, yellow-brown, red / white	adamantine / transpar., transluc.	6.5 / 4.2 — 4.3	good / brittle	acicular to capillary crystals; reticulated agg.; inclusions in rock crystal / tetragonal	as fine needles in alpine cracks, inclusions in quartz / quartz / cf. No. 541	Rutile variety **SAGENITE** TiO_2 / Brazil
348	yellow, red, red-brown, black / yellow, brown	adamantine, metal-like / transpar., transluc.	5.5 — 6 / 3.9 — 4.2	poor / uneven	thin-tabular, flat crystals / orthorhombic	in alpine cracks / anatase, titanite, adularia, quartz / rutile, cf. No. 198	**BROOKITE** TiO_2 / Riedertobel near Amsteg, Switzerland
349	dark yellow, brown, red, green, yellowish-white / white, grey	vitreous, greasy / transpar., transluc.	6 — 6.5 / 3.1 — 3.3	clear / conchoidal	small, short-columnar crystals; disseminated, rounded grains, compact, granular / monoclinic	in metamorphic limestones, in volcanic ejecta, in ore deps. / magnetite, galena & o. / olivine, cancrinite	**CHONDRODITE** $Mg_5[(OH),F]_2$ $(SiO_4)_2$ / Franklin, New Jersey, USA
350	brown, blackish-red, red, red-brown / red	vitreous / transluc., opaque	6.5 / 3.4	good / uneven, brittle	columnar to tubular crystals; long-radiating agg.; compact, radial, granular to massive / monoclinic	in crystalline schists, in manganese ore deps. / — / —	**PIEDMONTITE** $Ca_2(Mn,Fe)Al_2$ $[O\ OH\ SiO_4$ $Si_2O_7]$ (Manganiferous Epidote) / Aosta, Piedmont, Italy
351	brown, green, blue, yellow, red / white, grey	vitreous, greasy / transluc.	6.5 / 3.2 — 3.4	— / uneven & splintery	columnar, needle-shaped, pyramidal, thick-tabular crystals; compact, granular, massive, radial, tubular / tetragonal	in metamorphic rocks, in crystalline schists, in volcanic ejecta / chlorite, diopside, garnet, epidote / grossular, rutile, zircon, epidote	**VESUVIANITE** $Ca_{10}(Mg,Fe)_2Al$ $[(OH)_4(SiO_4)_5$ $(Si_2O_7)_2]$ / Monte Somma, Vesuvius, Italy
352	brown / white, grey	vitreous / transluc., opaque	6.5 / 3.2 — 3.4	— / uneven & splintery	long-columnar, striated crystals; radial, tubular agg. / tetragonal	radial vesuvianite	Vesuvianite variety **EGERAN** $Ca_{10}(Mg,Fe)_2Al$ $[(OH)_4(SiO_4)_5$ $(Si_2O_7)_2]$ / Hazlov, Czechoslovakia

No.	Colour Streak	Lustre Transparency	Hardness Specific Gravity	Cleavage Fracture & o. Phys. Props.	Common Form, Aggregates Crystalline Syst.	Occurrence Assoc. Minerals Similar Minerals	Name & Chem. Formula Origin of Specimen
353	green-brown, brown, grey-green / white, grey	vitreous, greasy transluc., opaque	6.5 / 3.4	— / uneven, splintery	short-columnar, well-formed crystals / tetragonal	in metamorphic tuffaceous rocks, embedded / grossularite / —	Vesuvianite variety **WILUITE** $Ca_{10}(Mg,Fe)_2$ $Al_4[(OH)_4$ $(SiO_4)_5 (Si_2O_7)_2]$ / Vilyui River, Yakutia, USSR
354	yellow-green, red-brown, greenish-black, olive-green / white	vitreous, greasy transpar., transluc.	6.5 — 7 / 3.3	good / conchoidal	tabular, columnar crystals; granular, massive, compact, insets / orthorhombic	constituent in basic igneous rocks, in magnetite deps. / — / cf. No. 299	**OLIVINE** $(Mg,Fe)_2[SiO_4]$ (Peridote, Chrysolite) / Kraslice, Czechoslovakia
355	colourless, yellow, green, brown, red, blue, orange / white	vitreous, adamantine, greasy transpar., transluc.	6.5 — 7.5 / 3.9 — 4.8	imperfect / conchoidal	columnar, pyramidal, needle-shaped, embedded crystals; granular, botryoidal / tetragonal	in igneous rocks, in fissures, in crystalline schists / biotite, amphibole, garnet, quartz & o. / vesuvianite, thorite, titanite, garnet	**ZIRCON** $Zr[SiO_4]$ / Ilmen Mountains, Southern Urals, USSR
356	brown, brown-green, green, black / white, grey	vitreous, greasy transluc., opaque	7 / 3.3 — 4.1	— / conchoidal, brittle	dodecahedral, icosi-tetrahedral crystals; compact, granular, massive / cubic	in crystalline schists & metamorphic rocks, in ore deps. / — / vesuvianite, zircon, ruby	**ANDRADITE** $Ca_3Fe_2[SiO_4]_3$ (Common Garnet) / Ciclova, Romania
357	brown, grey, violet, greenish / white	vitreous transpar., transluc.	6.5 — 7 / 3.3	good / conchoidal, brittle	tabular, wedge-shaped & lentiform crystals; granular, compact, disseminated / triclinic	in granites in fissures, in crystalline schists, in metamorphic limestone, in ore deps. / epidote, chlorite & o. / titanite	**AXINITE** $Ca_2(Fe,Mn)Al_2$ $[BO_3OH Si_4O_{12}]$ / Bourg d'Oisans, Dauphiné, France
358	brown, brownish-black / grey-white	vitreous, dull transluc., opaque	7 — 7.5 / 3.7 — 3.8	clear / conchoidal, splintery	short- & long-columnar crystals; often cruciform twins / orthorhombic	in crystalline schists / kyanite, almandine, muscovite, quartz / garnet	**STAUROLITE** $Al_4[Fe(OH)_2O_2$ $(SiO_4)_2]$ / Quimper, France
359	brown, brown-black / white	vitreous, greasy transpar., transluc.	7 — 7.5 / 3 — 3.2	— / conchoidal, brittle	long-columnar, longitudinally striated, embedded crystals / trigonal	in crystalline schists, in metamorphic limestone / — / ilvaite	Tourmaline variety **DRAVITE** $NaMg_3Al_6$ $[(OH)_4 (BO_3)_3$ $Si_6O_{18}]$ (Tourmaline Family) / Dobrova, Yugoslavia
360	brown, red, yellow, black / white	vitreous transpar., transluc.	8 / 3.5 — 4.1	— / conchoidal	octahedral embedded crystals; granular / cubic	in metamorphic limestone & dolomite, in basalts, in volcanic ejecta / — / corundum, zircon, garnet, cf. No. 246	**SPINEL** Al_2MgO_4 / Amity, New York, USA

No.	Colour / Streak	Lustre / Transparency	Hardness / Specific Gravity	Cleavage / Fracture & o. / Phys. Props.	Common Form, Aggregates / Crystalline Syst.	Occurrence / Assoc. Minerals / Similar Minerals	Name & Chem. Formula / Origin of Specimen
361	lead-grey, blackish-grey / grey, black	metallic / opaque	1 – 1.5 / 6.8 – 7.5	very good / flexible, soft, malleable	thin-tabular, lamellar crystals; scaly agg.; compact, granular / orthorhombic	in gold veins / sylvanite, gold, pyrite, tellurium, quartz / —	NAGYAGITE $AuTe_2 . 6Pb(S,Te)$ (Black Tellurium) / Săcărâmb, Romania
362	lead-grey to bluish / dark grey, leek-green, shining	metallic / opaque	1 – 1.5 / 4.6 – 4.7	perfect / inelastic, flexible, soft, greasy	flat, tabular crystals; lamellar, fine-scaly, compact to massive, disseminated / hexagonal	in granites, pegmatites, in tin ore deps., in quartz veins / cassiterite, wolframite, fluorite, tourmaline, apatite / graphite	MOLYB-DENITE MoS_2 (Molybdic Ochre) / Krupka, Czechoslovakia

No.	Colour / Streak	Lustre / Transparency	Hardness / Specific Gravity	Cleavage / Fracture & o. / Phys. Props.	Common Form, Aggregates / Crystalline Syst.	Occurrence / Assoc. Minerals / Similar Minerals	Name & Chem. Formula / Origin of Specimen
363	silver-white, grey to blackish / silver-white	metallic / opaque	2.5 – 3 / 10.1 to 11.1	hackly / ductile, malleable	filiform, sheet- & curl-shaped, mossy, compact / cubic	in ore veins / argentite, pyrargyrite, galena & o. / — cf. No. 5, 6	nat. SILVER Ag / Wittichen, Black Forest, Germany
364	tin-white, grey / grey	metallic / opaque	2 – 2.5 / 6.1 – 6.3	very good / sectile but brittle	columnar, short-needle-shaped crystals; mostly compact, finely granular, fine-tubular, disseminated / trigonal	associated with nat. gold / sylvanite, pyrite, quartz & o. / —	nat. TELLURIUM Te / Fata Bai, Romania
365	dark lead-grey, black / dark grey, shining	metallic, dull / opaque	2 – 2.5 / 6.1 – 6.3	indistinct / hackly, sectile, malleable, flexible	cubic, octahedral, dodecahedral crystals; disseminated, platy, powdery, sooty, reticulated, compact / monoclinic	in pseudomorphs after silver, in silver ore deps. / silver, pyrargyrite, proustite, galena & o. / chalcosine	ARGENTITE Ag_2S (Silver Glance) / Freiberg, Ore Mountains, Germany
366	lead-grey / grey-black, dull	metallic, dull / opaque	2.5 / 7.2 – 7.6	perfect, cube / spathic, brittle to soft	cubo-octahedral crystals; compact, granular, massive, disseminated, reticulated, platy / cubic	in hydrothermal ore veins, metasomatic in limestone & dolomite, sedimentary / sphalerite, chalcopyrite, tetrahedrite, pyrite, antimonite, bournonite	GALENA PbS (Lead Glance, Blue Lead) / Joplin, Missouri, USA
367	lead grey to black / black, dark grey	metallic / opaque	2 – 2.5 / 4.6 – 4.7	perfect / soft, inelastic, flexible, conchoidal	long-columnar, needle-shaped, fragile, capillary, often longitudinally striated crystals; compact, tubular, fibrous, radiate in bunches, capillary entangled, granular, massive / orthorhombic	in ore veins, in quartz veins, metasomatic in limestones / barytes, siderite, pyrite, nat. gold, realgar, orpiment, quartz & o. / bismutite, pyrolusite, galena	ANTIMONITE Sb_2S_3 (Antimony Glance, Grey Antimony, Stibnite) / Ichinokawa Mine, Japan
368							ANTIMONITE Sb_2S_3 / Capnic, Romania

No.	Colour / Streak	Lustre / Transparency	Hardness / Specific Gravity	Cleavage / Fracture & o. / Phys. Props.	Common Form, Aggregates / Crystalline Syst.	Occurrence / Assoc. Minerals / Similar Minerals	Name & Chem. Formula / Origin of Specimen
369	lead-grey, steel-grey / light grey, shining	metallic / opaque	2.5 – 3 / 8.1 – 8.8	indistinct / brittle, sectile	distorted, cube-shaped crystals; mostly compact, granular / monoclinic >155° cubic	in hydrothermal ore deps. / nat. gold, tellurium, nagyagite, pyrite, sylvanite & o. / —	**HESSITE** Ag_2Te (Silver Telluride) / Botesti, Romania
370	white-grey, dark grey / dark grey, shining	metallic / opaque	2.5 – 3 / 5.5 – 5.8	indistinct / conchoidal, soft	thick-tabular, short-columnar, needle-shaped crystals; compact, massive, finely granular, disseminated, earthy, encrustations / orthorhombic >103° hexagonal	in hydrothermal ore veins, also sedimentary in sandstone & tufa / chalcopyrite, bornite, covellite, pyrite & o. / tetrahedrite, bornite, argentite	**CHALCOCITE** Cu_2S (Copper Glance) / Redruth, Cornwall, England
371	dark reddish-grey, brown-red, violet / black	metallic / opaque	2.5 – 3 / 4.5	— / brittle	only compact, finely granular, no crystals / cubic	rare in ore veins / sphalerite, pyrite, tetrahedrite, bornite / bornite, enargite	**GERMANITE** $Cu_3(Fe,Ge)S_4$ / Tsumeb, South-West Africa
372	lead-grey to black / dark grey	metallic / opaque	2.5 – 3 / 7.6 – 7.8	very good / soft	no crystals; only compact, finely granular / cubic	in hydrothermal ore deps. / nat. gold, haematite, calcite, dolomite, barytes / galena	**CLAUS-THALITE** PbSe / Falun, Kopparberg, Svealand, Sweden
373	blackish-grey to black / black, shining	metallic / opaque	2.5 / 6.2 – 6.3	poor / soft to brittle, conchoidal	thick-tabular, columnar crystals; rose-shaped clusters, compact, disseminated, encrustations / orthorhombic	in silver veins / silver, argentite, baryte, fluorite & o. / argentite, chalcocite	**STEPHANITE** $5Ag_2S . Sb_2S_3$ (Brittle Silver Ore) / Jáchymov, Czechoslovakia
374	blackish-grey / dark grey, black	metallic / opaque	2.5 – 3 / 5.7	clear / soft	pyramidal, needle-shaped, striated crystals; often distorted, needle-shaped, fibrous, satiny agg. / monoclinic	in lead-zinc ore deps. / — / jamesonite, boulangerite	**HETERO-MORPHITE** $11PbS . 6Sb_2S_3$ / Příbram, Czechoslovakia
375	steel-grey, dark grey / grey-black	metallic / opaque	2 – 2.5 / 5.5 – 5.7	clear / soft	long-acicular, fine-capillary, fragile crystals; radial-fibrous, bunchy, entangled agg.; compact, massive, feather-like / monoclinic	in hydrothermal ore veins / galena, bournonite & o. / boulangerite, heteromorphite, antimonite	**JAMESONITE** $4PbS . FeS . 3Sb_2S_3$ (Feather Ore) / Cornwall, England
376	grey-black / black	metallic / opaque	2.5 / 5.9 – 6.1	clear / brittle	tabular prismatic crystals; radial-fibrous, globular agg.; crusts / monoclinic	in lead ore deps. / galena, sphalerite & o. / —	**SEMSEYITE** $9PbS . 4Sb_2S_3$ / Herja (Chiuzbaia, Kisbánya), Romania

No.	Colour / Streak	Lustre / Transparency	Hardness / Specific Gravity	Cleavage / Fracture & o. Phys. Props.	Common Form, Aggregates / Crystalline Syst.	Occurrence / Assoc. Minerals / Similar Minerals	Name & Chem. Formula / Origin of Specimen
377	blackish-grey / black	metallic / opaque	2.5 / 5.4	— / lamellar, divisible, soft to brittle, malleable	lamellar-cylindrical, shelly agg.; cylindrical bunches, spheroidal / orthorhombic	in hydrothermal silver-tin veins / cassiterite, stannine, sphalerite, pyrite, jamesonite	**CYLINDRITE** $Pb_3Sn_4Sb_2S_{14}$ / Poopó, Bolivia
378	lead-grey, blackish-grey / black	metallic, dull, silky / opaque	2.5 / 5.8−6.2	clear / flexible, brittle	long-columnar, needle-shaped, striated crystals; fibrous, radial, feather-like agg.; compact, massive / monoclinic	in hydrothermal ore veins / antimonite, galena, sphalerite, arsenopyrite, pyrite & o. / jamesonite, heteromorphite	**BOULANGERITE** $5PbS . 2Sb_2S_3$ / Příbram, Czechoslovakia
379	grey, grey-black / black	metallic, sub-metallic, indistinct / opaque	2−7 / 5	— / brittle	radial, fibrous, needle-shaped, capillary, compact, radiate crystal masses; massive, earthy, reniform, cone-shaped, loose / tetragonal	in oxidized zone of manganese ores, hydrothermal in ore veins, sedimentary deps. / psilomelane, manganite, braunite / —	**PYROLUSITE** MnO_2 / Příbram, Czechoslovakia
380	tin-white, lead-grey, black-tarnished / black	metallic, dull / opaque	3 / 5.7	perfect / brittle, shelly	mostly of crooked shelly form, reniform, compact masses; shelled refraction, mammillated, nodular, platy, compact / trigonal	in ore veins / silver, galena, proustite, sphalerite, chloanthite, niccolite, fluorite, baryte & o.	**nat. ARSENIC** As / St. Andreasberg, Harz, Germany
381	tin-white to lead-grey / lead-grey	metallic / opaque	3.5 / 6.6	perfect / brittle	tabular, cube-like crystals; mostly compact, granular-spathic, lamellar, reniform, disseminated / trigonal	in ore veins / antimonite, allemontite & o. / allemontite, dyscrasite	**nat. ANTIMONY** Sb / Allemont, France
382	lead-grey / grey-black	metallic / opaque	3−4 / 4.5−5	— / uneven, brittle	tetrahedral, cubic, dodecahedral crystals; compact, granular, disseminated, massive / cubic	in hydrothermal silver veins / galena, chalcopyrite, sphalerite, pyrite, argentite / bournonite, tetrahedrite	**FREIBERGITE** $(Cu,Ag)_3$ $(Sb,As)S_4$ (Silver Fahlerz, Argentiferous Grey Copper Ore) / Freiberg, Germany
383	lead-grey, black / grey	metallic, greasy / opaque	3 / 5.7−5.9	indistinct / brittle to soft, conchoidal	tabular, short-columnar, cube-shaped, pyramidal crystals; interpenetrant twins, compact, granular, massive, disseminated, encrustations / orthorhombic	in hydrothermal ore veins / galena, sphalerite, chalcopyrite, tetrahedrite, antimonite / tetrahedrite, boulangerite	**BOURNONITE** $2PbS . Cu_2S . Sb_2S_3$ (Wheel Ore, Endellionite) / Neudorf near Quedlinburg, Harz, Germany
384	grey, yellowish, tarnishing to bronze colour / black	metallic / opaque	4 / 4.3−4.5	indistinct / brittle, conchoidal	pseudocubic, tetrahedral crystals; compact, finely granular, disseminated, massive / tetragonal	in ore veins / cassiterite, tetrahedrite, arsenopyrite, pyrite / chalcopyrite, tetrahedrite, arsenopyrite	**STANNINE** Cu_2FeSnS_4 (Tin Pyrites, Bell Metal Ore) / Oruro, Bolivia

378

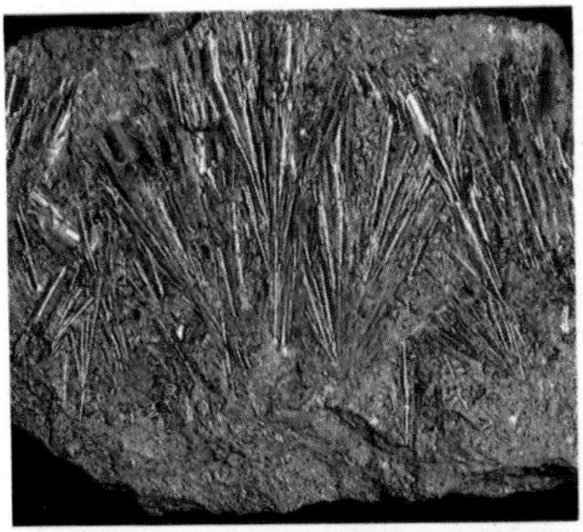

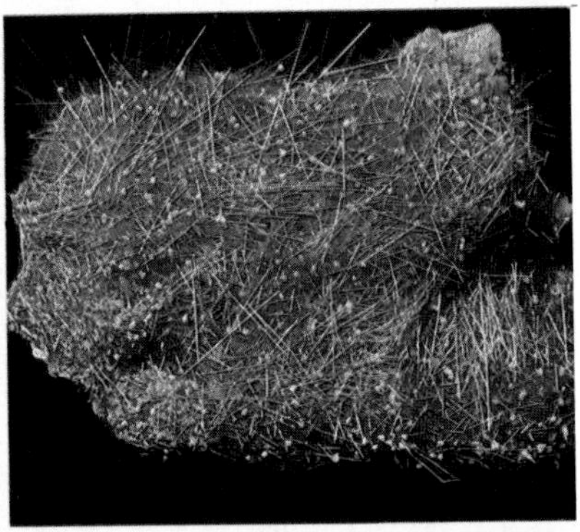

380

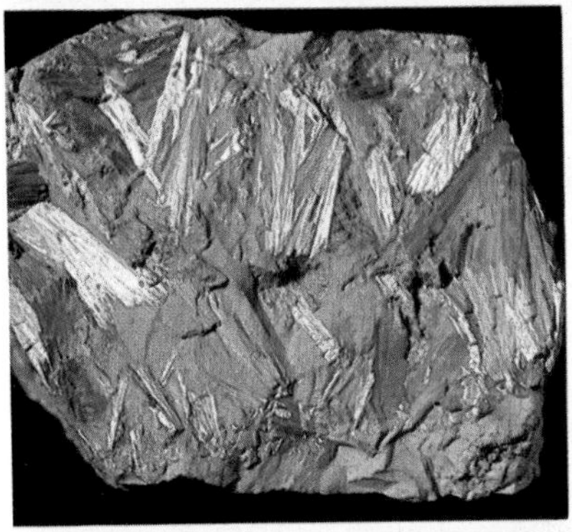

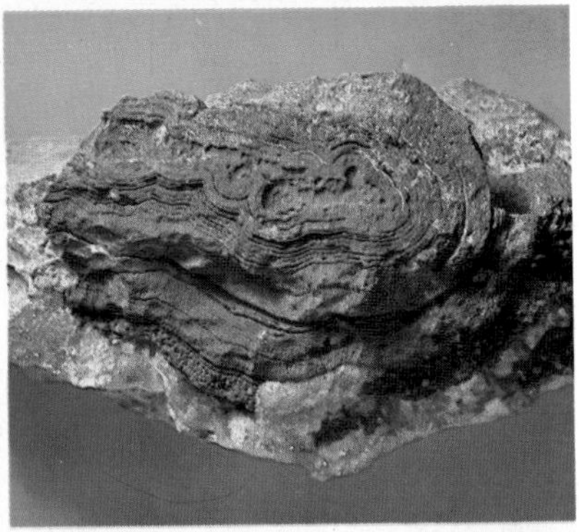

382

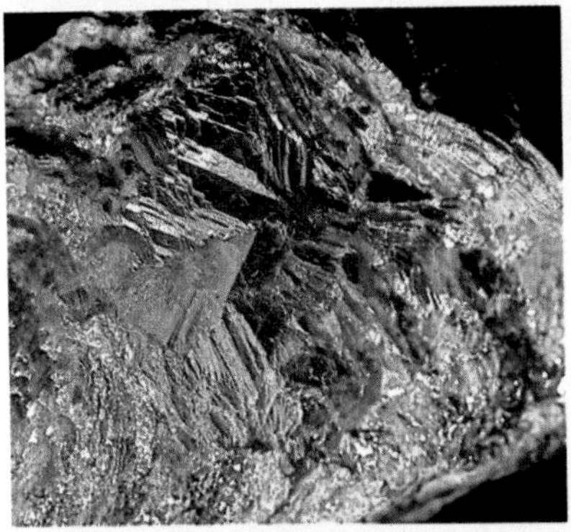

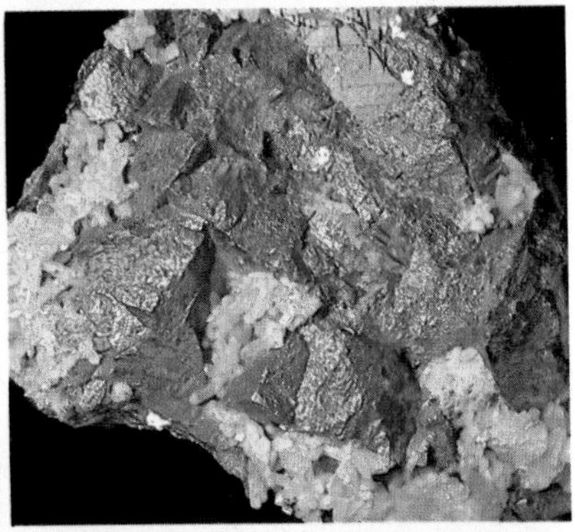

384

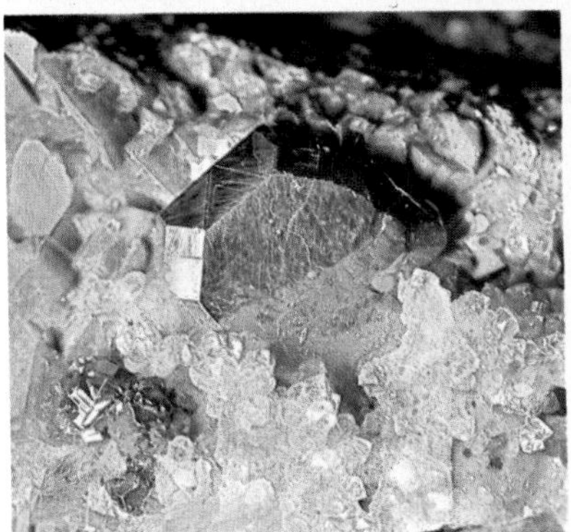

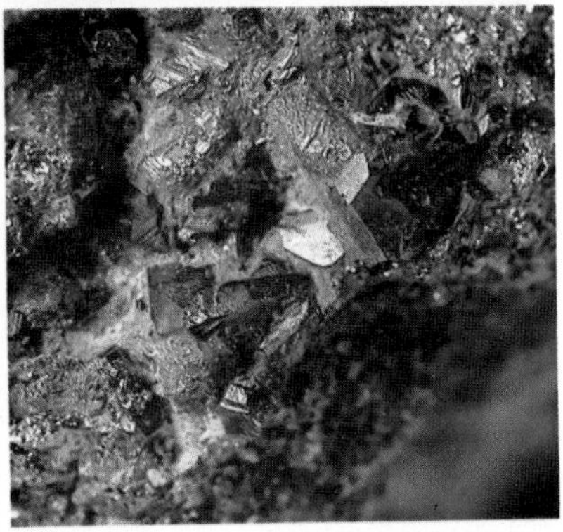

No.	Colour / Streak	Lustre / Transparency	Hardness / Specific Gravity	Cleavage / Fracture & o. / Phys. Props.	Common Form, Aggregates / Crystalline Syst.	Occurrence / Assoc. Minerals / Similar Minerals	Name & Chem. Formula / Origin of Specimen
385	steel-grey / grey, shining	metallic opaque	4—5 / 7.3—7.8	good / hackly, malleable, ductile	compact, disseminated, in grains, scales, clumps / cubic	in all meteorites, rarely as inclusions in basalt, in ultrabasic rocks / pentlandite / —	nat. IRON / Fe / Blaafjeld near Ovifak, Greenland
386	copper-red, grey, tarnishing to blue / reddish-brown, black	metallic opaque	5.5 / 8.2	poor / uneven, brittle	thin-tabular, columnar, needle-shaped crystals; compact, finely granular, lamellar, disseminated / hexagonal	in ore veins / chloanthite, ullmannite, pyrargyrite, galena & o. / niccolite	BREIT-HAUPTITE / NiSb / Monte Narba, Sardinia
387	reddish-silver-white, tarnishing to steel-grey / grey-black	metallic opaque	4.5—5.5 / 4.8	changing / uneven, brittle	octahedral, cubic crystals; compact, granular, disseminated / cubic	in hydrothermal ore veins / galena, tetrahedrite, chalcopyrite / ullmannite, gersdorffite, cobaltite	LINNAEITE / Co₃S₄ / Schwaben Min, Müsen, Siegerland, Germany
388	tin-white, grey-white / black, grey	metallic opaque	4.5—5 / 7.2—7.4	indistinct / conchoidal, brittle	lentiform, disc-shaped crystals; stellate triplets, compact, reniform, botryoidal, finely granular, radiate, fibrous, massive / orthorhombic	in ore veins / niccolite, bismuth, löllingite, siderite, barytes, fluorite / chloanthite, skutterudite, arsenopyrite	SAFFLORIT) / CoAs₂ (Smaltite, Tin White Cobalt) / Schneeberg, Germany
389	tin-white, grey / black	metallic opaque	5.5—6 / 6.2—6.9	clear / brittle	short-columnar, tabular crystals; compact, granular, radial, fibrous / orthorhombic	in ore veins / silver, galena & o. / löllingite, chloanthite, skutterudite, cf. No. 15	ARSENO-PYRITE / FeAsS (Mispickel, Arsenical Pyrites) / Ciclova, Romania
390	tin-white to dark grey / grey-black	metallic opaque	5.5 / 6.6	— / jointing due to zoning, brittle, fibrous	cubic, often ventricular, contorted crystals; finely granular, compact, reniform, massive / cubic	in hydrothermal ore veins / arsenic, bismuth, galena, niccolite, ullmannite & o. / safflorite, löllingite, arsenopyrite, rammelsbergite, cf. No. 14	CHLOAN-THITE / NiAs₃ (White Nickel) / Schneeberg, Germany
391	tin-white, tarnishing dark grey / grey-black	metallic opaque	5.5 / 6.5	— / uneven, brittle	octahedral crystals; compact, granular, massive, reniform, reticulated / cubic	in silver-cobalt-nickel veins / silver & similar minerals / safflorite, cobaltite, löllingite, arsenopyrite, ullmannite, rammelsbergite, cf. No. 16	SKUTTE-RUDITE / CoAs₃ (Smaltite) / Huelva on Gulf Cadiz, Spain
392	tin-white, tarnishing grey / grey-black	metallic opaque	5.5—6 / 7.1	clear / brittle	lentiform, prismatic crystals; compact, finely radiate, granular / orthorhombic	in ore veins / smaltite, niccolite, chloanthite, safflorite, arsenopyrite, löllingite, skutterudite	RAMMELS-BERGITE / NiAs₂ / Richelsdorf, Hessen, Germany

No.	Colour / Streak	Lustre / Transparency	Hardness / Specific Gravity	Cleavage Fracture & o. / Phys. Props.	Common Form, Aggregates / Crystalline Syst.	Occurrence / Assoc. Minerals / Similar Minerals	Name & Chem. Formula / Origin of Specimen
393	white, grey, yellowish, brown / white, yellowish	adamantine / transpar., transluc.	1−2 / 6.4−6.5	good / soft, sectile	tabular, columnar, short-acicular crystals; compact, crusty, earthy, carneous, massive, encrustations / tetragonal	in quicksilver ore deps. / quicksilver, cinnabar, barytes, quartz / —	**CALOMEL** Hg_2Cl_2 (Horn Quicksilver) / Moschellandsberg, Rhine-Palatinate, Germany
394	yellow-brown, grey, white-grey / white	vitreous, dull / transluc.	2 / 1.6−1.7	good / soft	roof-like, coffin-shaped, tabular, only embedded crystals / orthorhombic	in manure pits & canals, in guano / — / —	**STRUVITE** $NH_4Mg[PO_4] \cdot 6H_2O$ / Hamburg, Germany
395	white, grey, grey-rose / white	pearly / transluc.	3−3.5 / 3	very good / brittle, friable	thin-tabular, lamellar, pseudo-hexagonal crystals; compact, scaly, granular-lamellar, disseminated / monoclinic	in chlorite schists, in emery deps. / corundum, magnetite, rutile, emerald / —	**MARGARITE** $CaAl_2[(OH)_2 Al_2Si_2O_{10}]$ / Zillertal, Tyrol, Austria
396	white, grey, yellowish, brownish / white	dull / opaque	1−2 / 2.5−2.6	— / soft, swelling in water	loosely compact, clay-like; massive, earthy, friable masses / monoclinic	clay composed of montmorillonite, sedimentary / — / kaolinite, pyrophyllite	**BENTONITE** $Al_2Mg(OH)_2 [Si_4O_{10}] (Ca,Na)_x \cdot 4H_2O$ / Rock River, Wyoming, USA
397	white, yellow, grey-white / white	pearly, silver glance / transluc.	1 / 2.5−2.6	very good / soft, flexible, greasy	pseudohexagonal, thin-foliaceous crystals; fine-scaly, compact, massive, earthy, loose / monoclinic	hydrothermal in tin ore deps., in ore veins / — / pyrophyllite, sericite, talc	**NACRITE** $Al_4[(OH)_8 Si_4O_{10}]$ / Brand-Erbisdorf, Germany
398	yellowish, white, grey, greenish / white	silky, vitreous / transluc.	1−2.5 / 2.6	good / soft to brittle	compact masses, radiate-lamellar, globular, stellate, concentrically shelled agg.; granular / monoclinic	constituents: partly nacrite & kaolinite, partly muscovite, also sericite, pseudomorphoses after topaz, in tin ore deps., in granite / fluorite / —	**GILBERTITE** $Al_4[(OH)_8 Si_4O_{10}], KAl_2 [(OH,F)_2 AlSi_3O_{10}]$ / St. Just, Cornwall, England
399	grey, white, yellowish, brownish / white	dull / transluc., opaque	1−2 / 2.1−2.3	— / soft, flexible	finely fibrous, felt-like, cryptocrystalline / monoclinic or orthorhombic	weathering product of serpentine / chalcedony, opal, chlorite, magnesite / —	**PALYGOR-SKITE** $(Mg,Al)_2 [OH Si_4O_{10}] \cdot 4H_2O$ / Lovinobaňa, Czechoslovakia
400	olive-green, blackish-green, grey-black / greenish-grey	dull, pearly / opaque	2.5 / 3.2	good / soft	compact, finely granular, massive, fine-scaly, oölitic, cryptocrystalline / monoclinic	in sedimentary ore deps., in chlorite schists / limonite, magnetite, garnet / chamosite	**THURINGITE** $(Fe_2,Fe_3,Al)_3 [(OH)_2Al_2Si_2O_{10} (Mg,Fe_2,Fe_3)_3 (O,OH)_8]$ / Šternberk, Czechoslovakia

3	394
5	396
7	398
9	400

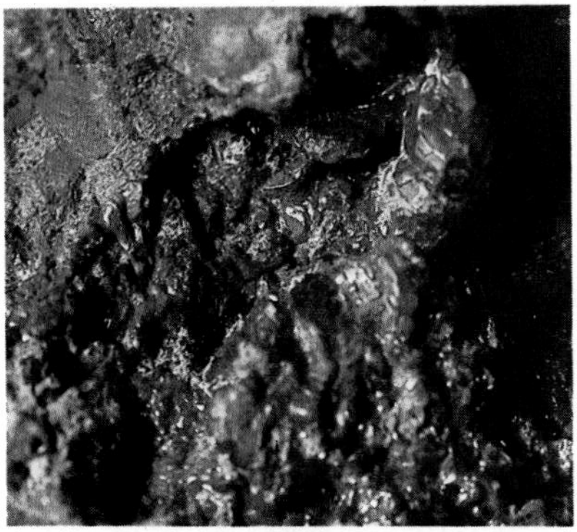

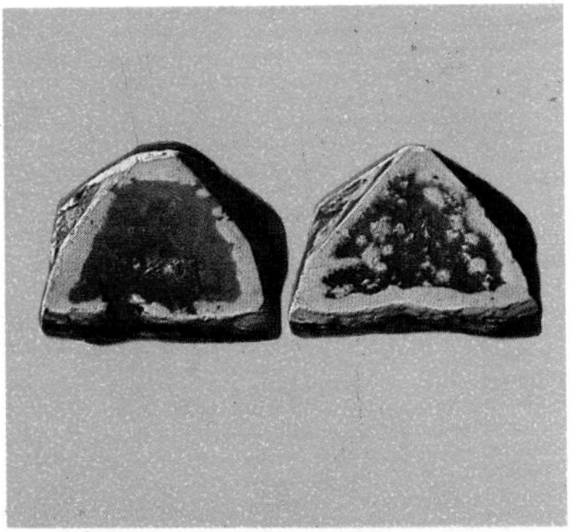

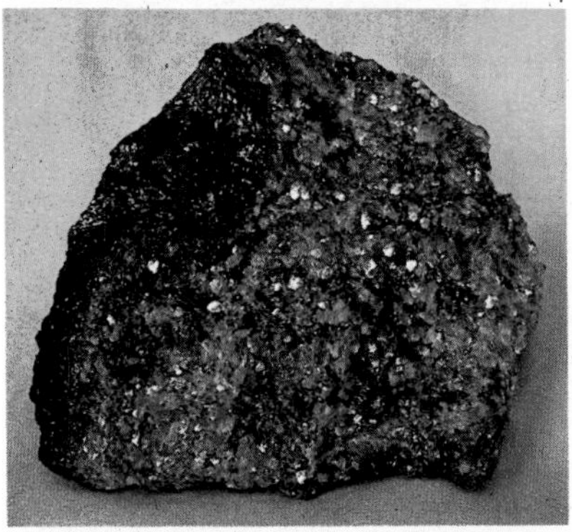

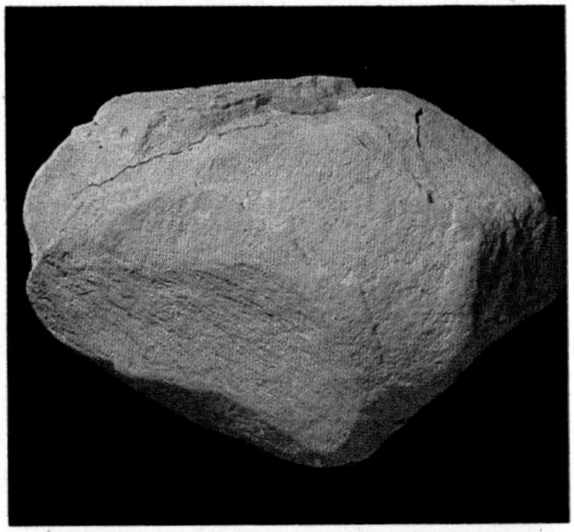

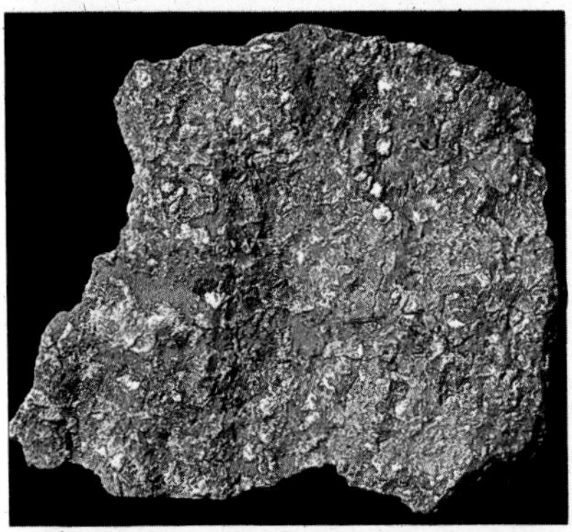

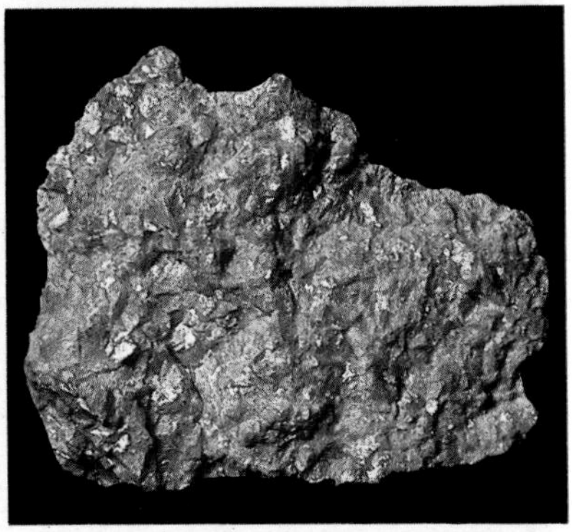

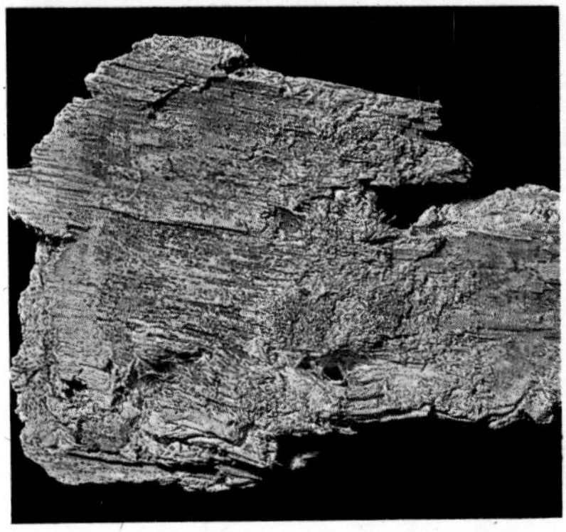

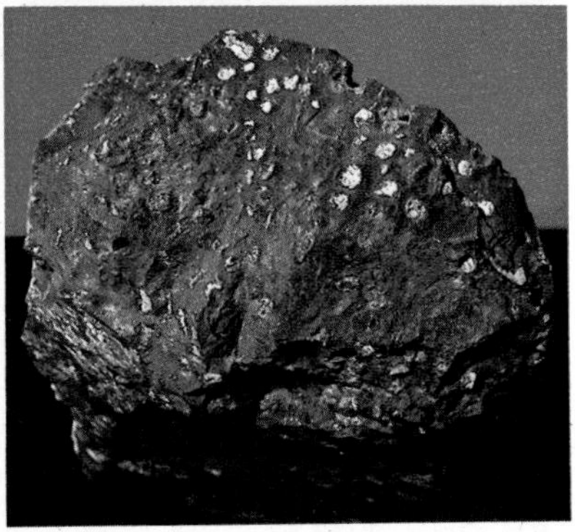

No.	Colour Streak	Lustre Transparency	Hardness / Specific Gravity	Cleavage Fracture & o. Phys. Props.	Common Form, Aggregates / Crystalline Syst.	Occurrence Assoc. Minerals / Similar Minerals	Name & Chem. Formula / Origin of Specimen
401	white, yellowish, grey, brown, black ___ white	adamantine, greasy transpar., transluc.	3 — 3.5 ___ 6.4 — 6.6	indistinct conchoidal, brittle	columnar, tabular, needle-shaped crystals; twins, triplets, stellate clusters, compact, reniform, massive ___ orthorhombic	in weathering zone of lead veins ___ — ___ anglesite, phosgenite, barytes, scheelite, cf. No. 321	CERUSSITE $PbCO_3$ (Ceruse, White Lead Ore) ___ Oloví, Czechoslovakia
402	colourless, white, grey, blackish, greenish ___ white	adamantine, greasy transluc., transpar.	3 ___ 6.1 — 6.4	good conchoidal, brittle	tabular, short-prismatic, columnar, needle-shaped crystals; compact, granular, reniform, crusty, earthy ___ orthorhombic	in oxidation zone of lead ore deps. ___ limonite, cerussite, galena ___ cerussite, barytes, scheelite	ANGLESITE $PbSO_4$ (Lead Vitriol) ___ Monte-Poni, Sardinia
403	brown, grey, yellowish, greenish ___ white	pearly, bronze-coloured transpar., transluc.	2.5 — 3 ___ 2.7 — 2.9	perfect flexible, foliaceous	pseudohexagonal, thin-tabular, short-prismatic crystals; coarse scales, lamellar agg. ___ monoclinic	in pegmatites, in metamorphic limestones & dolomites, in serpentine, in apatite veins ___ — ___ biotite, muscovite	PHLOGOPITE KMg_3 $[(F,OH)_2$ $AlSi_3O_{10}]$ ___ Ampandandrava, Madagascar
404	grey, greenish-grey, brownish, greenish-black ___ grey-green	vitreous, dull opaque	3 ___ 3	good uneven	massive, granular, compact, oölitic, concentrically shelled ___ monoclinic	sedimentary ore deps. ___ limonite, magnetite, siderite, calcite & o. ___ thuringite	CHAMOSITE (Fe_2Fe_3) $[(OH)_2 AlSi_3O_{10}]$ $(Fe,Mg)_3$ $(O,OH)_6$ ___ Nučice, Czechoslovakia
405	colourless, white, grey, yellowish, blackish ___ white	vitreous, dull transluc., transpar.	3.5 — 4 ___ 3	perfect conchoidal, brittle	rhombohedral crystals; compact, reniform, nodular, coarsely granular, spathic, massive ___ trigonal	metasomatic deps. replacing limestone & dolomite, in serpentine, in talc schists ___ — ___ ankerite, calcite, dolomite	MAGNESITE $MgCO_3$ (Bitter Spar) ___ Trieben, Styria, Austria
406	brown-red, grey to black, brownish-green ___ yellow-brown, orange-yellow	adamantine, vitreous, greasy, dull opaque	3.5 ___ 5.7 — 6.2	— conchoidal, brittle	columnar, pyramidal, tabular crystals; radiate agg.; crusty, botryoidal, mammillary, coatings ___ orthorhombic	in oxidation zone of ore veins ___ pyromorphite, vanadinite, cerussite, mimetesite, limonite ___ —	DESCLOIZITE $Pb(Zn,Cu)$ $[OH VO_4]$ ___ Total Wreck Mine, Arizona, USA
407	leek-green, grey, blackish ___ light grey	vitreous, greasy, dull transpar., transluc.	3.5 — 4 ___ 3.1 — 3.8	good splintery, brittle	short-columnar, pyramidal, tabular, needle-shaped crystals; globular agg.; reniform, finely granular, massive ___ orthorhombic	weathering product of iron-arsenic minerals, in cavities in limonite ___ arsenopyrite ___ pharmacosiderite	SCORODITE $Fe[AsO_4]$. $2H_2O$ ___ Graul Mine near Schwarzenberg, Germany
408	whitish-grey, yellowish-grey, brown ___ white	vitreous, pearly, silky transluc.	5 — 5.5 ___ 2.8 — 3.5	very good brittle	finely fibrous, fine needle-shaped, capillary, compact, nodular ___ orthorhombic	in metamorphic rocks, in mica schists ___ chrysotile, cf. No. 345	ANTHOPHYLLITE $(Mg,Fe)_7$ $[OH Si_4O_{11}]_2$ (Anthophyllite Asbestos) ___ Massachusetts, USA

No.	Colour Streak	Lustre Transparency	Hardness Specific Gravity	Cleavage Fracture & o. Phys. Props.	Common Form, Aggregates Crystalline Syst.	Occurrence Assoc. Minerals Similar Minerals	Name & Chem. Formula Origin of Specimen
409	yellow-brown, grey, black, irised tarnish — rusty brown, yellow	sub-vitreous, silky, greasy, dull — opaque	5—5.5 also softer — 3.6—3.7	— conchoidal, fibrous	cryptocrystalline, fibrous, tubular, reniform, cone-shaped, globular, shelled, stalactitic, oölitic, botryoidal, compact, earthy, loose, in pseudo-morphs after pyrite, siderite & o. ferruginous minerals orthorhombic	as weathering product in all iron deps., sedimentary deps., in hydro-thermal veins — pyrolusite, haema-tite, chalcedony & o. — haematite, cf. No. 482	LIMONITE Predominantly cryptocrystalline mixture of goe-hite (α-FeOOH) less often of lepidocrocite (γ-FeOOH) (Brown Iron O Brown Haema-tite, Brown Ironstone) Herdorf, Siegerland, Germany
410							LIMONITE Elba, Italy
411	white, yellow, red, green, grey, bluish — white	vitreous, greasy transluc.	5 — 4.3—4.5	very good brittle	small rhombo-hedral crystals; compact, botry-oidal, reniform, finely granular, massive, sinter crusts trigonal	metasomatic in limestone & dolo-mite, in zinc deps. — — hemimorphite, cf. No. 280	SMITH-SONITE $ZnCO_3$ (Galmei) Laurion, Greece
412	grey, greenish, brownish — white	vitreous, pearly transluc., opaque	5.5 — 3.1—3.3	clear uneven	columnar, thick-tabular, needle-shaped crystals; compact, granular, spathic orthorhombic	rock constituent in serpentine, in pegmatic apatite veins apatite, phlogopite, olivine, bronzite apatite, hypersthene	ENSTATITE $Mg_2[Si_2O_6]$ (Pyroxene Family) Bamle near Oedergården, Norway
413	brown, grey, green, black — white	vitreous, pearly, silky metallic — opaque	5—6 — 3.5—3.7	good striated cleavage planes	columnar, tabular, lamellar, granular, compact orthorhombic	rock constituent in basic rocks, in crystalline schists, in volcanic ejecta olivine, enstatite enstatite, diallage	HYPER-STHENE $(Fe,Mg)_2[Si_2O_6]$ St.Paul Island, Labrador
414	white, light grey, greenish — white	vitreous, silky transluc.	5—6 — 2.9—3	very good brittle	needle-shaped, fibrous, capillary, radial, compact, asbestos-like monoclinic	in metamorphic limestone & dolo-mite, in talc schists — chrysotile, pecto-lite, wollastonite	TREMOLITE $Ca_2Mg_5(OH,F)_2$ $[Si_4O_{11}]_2$ (Amphibole Family) Aosta, Pied-mont, Italy
415	grey, brown, yellowish, green, rose-red — white	vitreous, pearly, silky transluc., opaque	5.5—6 — 3.2—3.4	very good uneven	long-columnar, needle-shaped, tubular, fibrous, lamellar, radial orthorhombic	in metamorphic rocks, in crystalline schists, in granites, in ore veins — bronzite	GEDRITE $(Mg,Fe)_6Al_2$ $[OH(AlSi)$ $Si_3O_{11}]_2$ (Amphibole Family) Gèdres, Pyre-nees, France
416	grey, brown, white, blue, blackish — white	vitreous, greasy transpar., transluc.	5.5—6 — 2.2—2.4	very good con-choidal	columnar, dode-cahedral crystals; compact, granular cubic	in eruptive rocks; phonolites, basalts, in volcanic ejecta leucite, nepheline sodalite, hauyne	NOSEAN $Na_8[SO_4$ $(AlSiO_4)_6]$ Laacher See, Eifel, Germany

No.	Colour / Streak	Lustre / Transparency	Hardness / Specific Gravity	Cleavage / Fracture & o. / Phys. Props.	Common Form, Aggregates / Crystalline Syst.	Occurrence / Assoc. Minerals / Similar Minerals	Name & Chem. Formula / Origin of Specimen
417	grey, yellowish, reddish / white	vitreous, greasy / transluc.	6 / 3.8−4.3	good / conchoidal / splintery	columnar, needle-shaped crystals; compact, finely to coarsely granular / hexagonal	manganiferous willemite variety, in zinc deps. / zincite, franklinite / —	**TROOSTITE** $(Zn,Mn)_2[SiO_4]$ / Sterling Hill near Ogdensburg, New Jersey, USA
418	yellow, green, brown-red, grey to black / white	vitreous, metallic, brilliant / transluc., opaque	6.5 / 4.2	comparatively good / conchoidal	thick-tabular, short-columnar crystals; granular, massive / orthorhombic	in pegmatites, in tin ore veins, in obsidians / — / —	**FAYALITE** $Fe_2[SiO_4]$ (Olivine Family) / Fayal Island, Azores
419	colourless, white, greyish / white	vitreous transpar., dull	6.5 / 2.9	— / conchoidal	rarely cube-shaped crystals; mostly compact, coarsely granular, irregular grains / cubic	in granite pegmatites / beryl, petalite / hyalite, quartz	**POLLUCITE** (Cs,Na) $[AlSi_2O_6]H_2O$ / San Piero in Campo, Elba, Italy
420	grey, brownish, reddish, yellowish / white	vitreous, pearly / transluc., opaque	6−6.5 / 2.5	very good / uneven, brittle	tabular, bench-shaped crystals; only embedded / monoclinic	constituent in eruptive rocks / quartz, biotite, amphibole, nepheline / orthoclase, plagioclase	**SANIDINE** $K[AlSi_3O_8]$ (Felspar Family) / Drachenfels, Pfälzer Wald, Germany
421	colourless, white, green-grey, grey, blue / white	vitreous, pearly / transpar., transluc.	5−6 / 2.7	clear / —	short-prismatic, thick-columnar, tubular crystals; granular, fibrous, radial, massive / tetragonal	in metamorphic rocks, as volcanic ejecta / — / felspars, spodumene, cf. No. 87	**MEIONITE** Ca_8 $[(Cl_2,SO_4,CO_3)_2$ $(Al_2Si_2O_8)_6]$ (Scapolite Family) / Vesuvius, Italy
422	grey, grey-bluish, grey-yellowish / white	vitreous, silky, dull / transluc.	6−7 / 2.5−2.6	— / conchoidal	fibrous to crypto-crystalline, reniform, globular, botryoidal, shelled, stalactitic, mammillary, crusts, nodules, concretions, amygdaloidal filling, fine-fibrous, finely radial / trigonal	micro- to crypto-crystalline quartz of gelatigenous origin, in fissures & cavities in basaltic rocks, in ore veins / quartz, zeolite, calcite / prehnite, cf. No. 501	**CHALCEDONY** SiO_2 / Tri Vody, Czechoslovakia
423							**CHALCEDONY** SiO_2 / Grákollur, Reydarfjördur, Iceland
424	grey, grey-brown to black / white	greasy, dull / transluc., opaque	7 / 2.5−2.6	— / conchoidal, splintery	nodular, platy concretions, often with white weathering crust, massive / amorphous	in sedimentary rocks, in chalk / — / —	Opal variety **FIRESTONE** $SiO_2 \cdot nH_2O$ (Flint) / Rügen Island

No. 417−424

Table 5

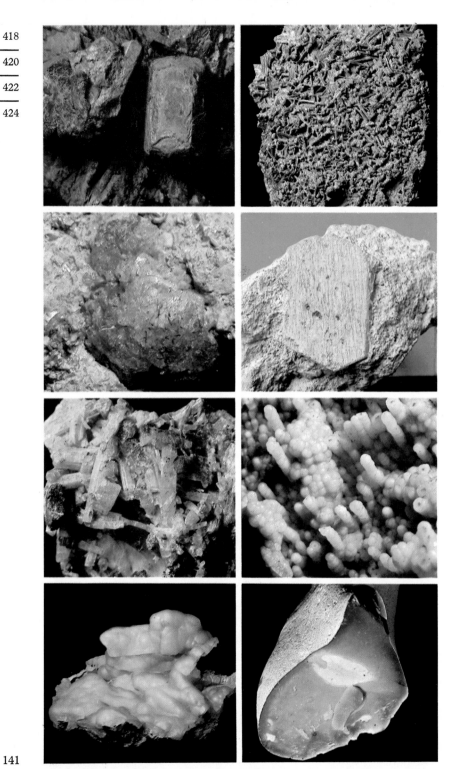

No.	Colour / Streak	Lustre / Transparency	Hardness / Specific Gravity	Cleavage / Fracture & o. / Phys. Props.	Common Form, Aggregates / Crystalline Syst.	Occurrence / Assoc. Minerals / Similar Minerals	Name & Chem. Formula / Origin of Specimen
425	steel-grey, black / black	sub-metallic shining, dull / opaque	$\frac{7}{4.8-5}$	clear / uneven, brittle	rhombohedral, small crystals; compact, finely crystalline, fibrous, granular, massive / tetragonal	idiomorphic pyrolusite, in manganese deps. / — / cf. No. 379	**POLIANITE** MnO_2 / Schneeberg, Germany
426	grey, reddish-grey, brownish / white	vitreous, greasy, dull / transluc., opaque	$\frac{7}{3.1-3.2}$	good to indistinct / uneven, brittle	thick-columnar to acicular crystals; granular, compact, tubular, radial, fibrous / orthorhombic	in metamorphic rocks, in mica schists & gneiss in pegmatites / garnet, tourmaline, rutile, quartz / tourmaline	**ANDALUSITE** $Al_2[O\ SiO_4]$ / Lisenz-Alpe, Sellrain, Tyrol, Austria
427	colourless, yellowish, dark brown, grey / white	vitreous, greasy / transpar., transluc.	$\frac{7-7.5}{2.9-3}$	indistinct / conchoidal	columnar crystals; granular, compact / orthorhombic	pneumatolytic in dolomite, in gneiss, in alpine cracks, in volcanic ejecta / — / topaz, datolite	**DANBURITE** $Ca[B_2Si_2O_8]$ / Russel, St.Lawrence Co., New York, USA
428	grey-brown, yellowish, reddish, bluish / white	vitreous, greasy, dull / transpar., transluc.	$\frac{9}{3.9-4.1}$	fissile (jointing) / conchoidal, splintery	columnar, barrel-shaped, thick-tabular crystals; granular, compact, massive / trigonal	in pegmatites, in crystalline schists, in metamorphic limestone, dolomite, syenite & basalt / zircon, spinel, apatite, topaz, cf. No. 512	**CORUNDUM** Al_2O_3 / Ambositra, Madagascar

No.	Colour / Streak	Lustre / Transparency	Hardness / Specific Gravity	Cleavage / Fracture & o. / Phys. Props.	Common Form, Aggregates / Crystalline Syst.	Occurrence / Assoc. Minerals / Similar Minerals	Name & Chem. Formula / Origin of Specimen
429	tombac-brown, red to black, iridescent tarnish / black	metallic / opaque	$\frac{3}{4.9-5.3}$	— / conchoidal, brittle to soft	cube-shaped, cubic octahedral, dodecahedral crystals; compact, platy masses, granular, massive, coatings / cubic	in copper veins / chalcopyrite, pyrite, chalcocite / pyrrhotine, niccolite (fresh), chalcocite, chalcopyrite, covellite	**BORNITE** Cu_5FeS_4 / Messina, Transvaal, South Africa
430	black / black	metallic / opaque	$\frac{3}{7.7-7.8}$	— / brittle	small, tetrahedral, dodecahedral crystals; powdery, coatings / cubic	in quicksilver deps. / cinnabar / —	**METACINNA-BARITE** HgS / Idrija, Slovenia, Yugoslavia
431	grey to black / black	metallic / opaque	$\frac{3-4}{4.5-5}$	— / conchoidal, brittle	tetrahedral, dodecahedral crystals; compact, granular to massive / cubic	in hydrothermal ore veins / galena, pyrite, bornite, chalcopyrite, sphalerite, siderite, quartz / chalcocite, bournonite, tennantite, cf. No. 483	**TETRA-HEDRITE** $(Cu,Zn,Ag,Fe)_3(Sb,As)S_{3-4}$ (Grey Copper, Fahlerz) / Capnic, Romania
432	dark grey, black, iridescent tarnish / black	metallic / opaque	$\frac{3}{6.4}$	very good / conchoidal	thin- to thick-tabular crystals; twin striation; globular, reniform masses / monoclinic	in ore veins, in dolomites / realgar, sphalerite, galena, pyrite & o. / —	**JORDANITE** $5\ PbS\ .\ As_2S_3$ / Binnental, Wallis, Switzerland

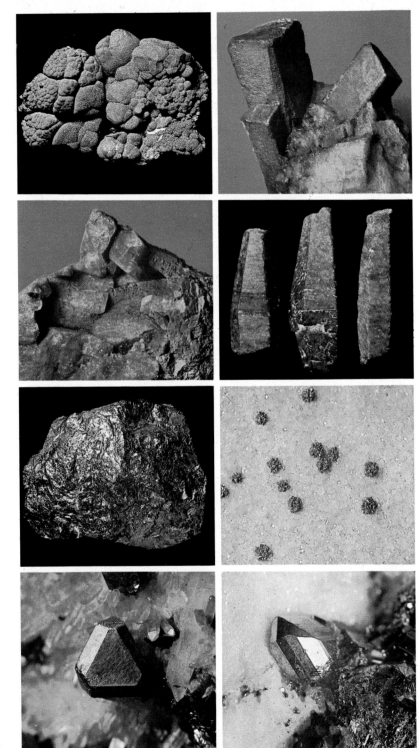

No.	Colour / Streak	Lustre / Transparency	Hardness / Specific Gravity	Cleavage / Fracture & o. / Phys. Props.	Common Form, Aggregates / Crystalline Syst.	Occurrence / Assoc. Minerals / Similar Minerals	Name & Chem. Formula / Origin of Specimen
433	black / black	metallic, greasy, dull / opaque	5.5 / 5.2	imperfect / con-choidal, brittle, magnetic	octahedral, dodeca-hedral crystals; compact, massive, coarsely & finely granular, dissemi-nated / cubic	isolated deps., in crystalline schists, in basalts, in alpine cracks / — / ilmenite, chromite, haematite	MAGNETITE Fe_3O_4 / Alto Adige, Italy
434	black / reddish-brown, brown	metallic, sub-metallic / opaque, transluc.	5.5 / 4.8 — 4.9	very good / uneven, brittle	pointed-pyramidal, pseudooctahedral crystals; compact, granular, spathic / tetragonal	in metasomatic & hydrothermal ore veins, in lime-stone & dolomite / pyrolusite, psilo-melane, braunite / magnetite, braunite, pyrolusite	HAUSMAN-NITE Mn_3O_4 / Ilmenau, Thuringia, Germany
435	black, steel-grey, reddish-brown, red, iridescent tarnish	metallic to dull / opaque	5.5 — 6.5 / 5.2 — 5.3	— / uneven, brittle, shelled separation	rhombohedral, tabular, lentiform crystals; lamellar, scaly agg.; rose-shaped clusters, compact, granular, radial, columnar, massive, earthy, mammillary / trigonal	in isolated deps., hydrothermal in ore veins, in meta-morphic rocks, metasomatic, sedi-mentary in alpine cracks, in craters of volcanoes & in lavas, as colouring / — / ilmenite, magnetite, chromite, frankli-nite, cf. No. 201	HAEMATITE Fe_2O_3 (Iron Glance, Specularite) / Rio Marina, Elba, Italy
436							HAEMATITE Fe_2O_3 / Gotthard area, Switzerland
437	cherry-red, reddish-brown						HAEMATITE Fe_2O_3 (Iron Mica) / Donovaly, Czechoslovakia
438	brown, black, grey, yellow / white, yellow, light brown	ada-mantine, metallic / transluc., opaque	6 — 7 / 6.8 — 7.1	indistinct / con-choidal, brittle	short-columnar, pointed-pyramidal, needle-shaped crys-tals; knee-shaped twins; compact, reniform, shelled, granular, fibrous, massive, mammil-lated masses / tetragonal	in pegmatites, in granites (gneiss) / wolframite, fluorite, topaz, zinnwaldite, scheelite, molybde-nite, apatite, quartz / tourmaline, sphale-rite, ilmenite, magne-tite, rutile, garnet	CASSITE-RITE SnO_2 (Tinstone) / Slavkov, Czechoslovakia
439							CASSITE-RITE SnO_2 (Tinstone) / Slavkov, Czechoslovakia
440	black / brownish-black	metallic, greasy / opaque	6 — 6.5 / 4.7 — 4.8	clear / brittle	octahedral crystals; compact, massive, granular, crusty / tetragonal	in hydrothermal, metasomatic veins, in limestone & dolomite, in ore deps. / hausmannite, pyrolusite, barytes, calcite, quartz / magnetite, haus-mannite, franklinite	BRAUNITE $MnMn_6$ $[O_8SiO_4]$ / Oehrenstock, Thuringia, Germany

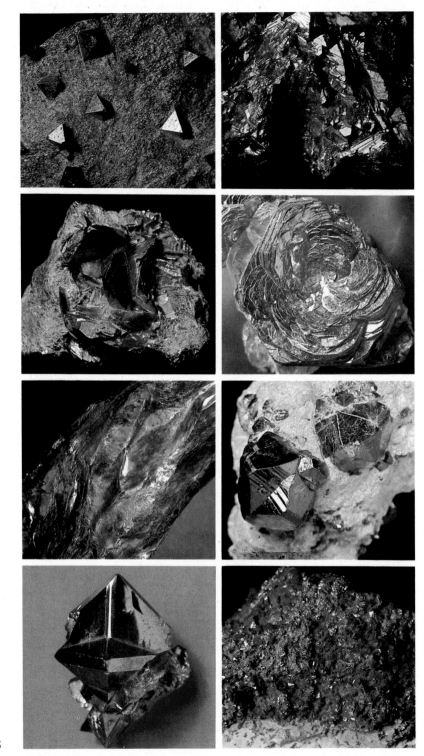

145

No.	Colour Streak	Lustre Transparency	Hardness Specific Gravity	Cleavage Fracture & o. Phys. Props.	Common Form, Aggregates Crystalline Syst.	Occurrence Assoc. Minerals Similar Minerals	Name & Chem. Formula Origin of Specimen
441	black, steel-grey black, grey, shining	sub-metallic, dull opaque	1−2 2−2.2	perfect flexible, greasy	thin-tabular crystals; lamellar, scaly, compact, massive, earthy, disseminated, dendritic hexagonal	in metamorphic rocks, crystalline schists, limestone, in pegmatites pyrite, garnet molybdenite	GRAPHITE C Ceylon
442	black, dark brown black, dark brown	dull opaque	1−2 2.3−3.7	— —	cryptocrystalline, compact, spumous, nodular, earthy, sooty, scaly, reniform, botryoidal, dendritic, coatings amorphous	in oxidized zone of manganese ore deps. psilomelane, pyrolusite, limonite asbolan	WAD $MnO_2 \cdot nH_2O$ Solnhofen, Fränkische Alb, Germany
443	black, grey, greenish brownish-black, dark green	greasy, dull, sub-metallic opaque	4−6 9−10.9	— conchoidal, brittle, radioactive	rarely cube-shaped, octahedral crystals; compact, reniform, globular, scaly, mammilliform, massive cubic	in granite- & syenite-pegmatites, in hydrothermal ore veins, sedimentary mica, galena, molybdenite, dolomite, fluorite thorianite, psilomelane	URANINITE UO_2 with (U,O_3,Th,Pb) (Pitch Blende, Pitch Ore) Příbram, Czechoslovakia
444	black, dark brown, greenish white, grey	vitreous, pearly, sub-metallic transpar., transluc.	2.5−3 2.8−3.2	perfect elastic, flexible, soft to brittle	tabular, columnar, pseudohexagonal crystals; compact, lamellar, scaly monoclinic	rock constituent in granites, pegmatites, basic rocks, basalts, metamorphic rocks, in volcanic ejecta vermiculite, phlogopite	BIOTITE $K(Mg,Fe,Mn)_3$ $[(OH,F)_2$ $AlSi_3O_{10}]$ (Iron Mica, Micaceous Iron Ore) Krupka, Czechoslovakia
445	red-yellow, grey, brown, green, black white, yellow, brown	adamantine, greasy, sub-metallic transpar., opaque	3.5−4 3.9−4.2	very good brittle	tetrahedral, dodecahedral, cube-shaped crystals; coarsely to finely granular, massive, fibrous cubic	in ore veins galena, chalcopyrite & o. cf. No. 142, 191, 315	SPHALERITE ZnS (Zinc Blende, Mock Ore, Mock Lead, Black Jack) Cumberland, England
446	grey to black grey-black, black	sub-metallic, dull opaque	3−4 5.1	— conchoidal, brittle	tetrahedral, dodecahedral, cube-shaped crystals; twins, compact, granular, disseminated, massive cubic	in hydrothermal ore veins cinnabar, quicksilver, chalcopyrite, barytes bournonite	SCHWAZITE $(CuHg)_3SbS_3$ (Mercury Fahlore) Kogel near Brixlegg, Tyrol, Austria
447	dark brown, black whitish, grey	vitreous, pearly, sub-metallic transluc.	3−4 3.3	very good uneven, brittle	small, prismatic crystals; granular orthorhombic	in metamorphic limestones spinel, magnetite, serpentine —	WAR-WICKITE $(Mg,Fe)_3Ti$ $[O BO_3]_2$ Ilfeld, Harz, Germany
448	black, greenish-black dark green	vitreous transluc., opaque	3.5−4 3.4	good brittle	hexagonal & sometimes triangular, pyramidal, tetrahedron-shaped crystals; globular clusters monoclinic, sometimes trigonal	in ore veins pyrite, siderite, calcite, sphalerite, limonite —	CRON-STEDTITE $Fe_4Fe_2[(OH)_8$ $Fe_2Si_2O_{10}]$ Příbram, Czechoslovakia

No.	Colour / Streak	Lustre / Transparency	Hardness / Specific Gravity	Cleavage / Fracture & o. / Phys. Props.	Common Form, Aggregates / Crystalline Syst.	Occurrence / Assoc. Minerals / Similar Minerals	Name & Chem. Formula / Origin of Specimen
449	black, dark brown, blackish-green	sub-metallic opaque	$\dfrac{3.5-4}{3.9-4}$	very good / brittle	small, octahedral or cube-shaped crystals; compact, granular, lamellar, disseminated / cubic	in ore veins / galena, sphalerite, pyrite, rhodonite, rhodochrosite / –	ALABAN-DITE MnS (Mangan-Blende) / Săcărâmb, Romania
450	black brown-black	sub-metallic, dull opaque	$\dfrac{4.5-5}{5.4}$	– / brittle	mammilliform, reniform, botry-oidal masses; concentrically shelled agg. / tetragonal	in weathering zone of manganese ore deps. / – / pyrolusite	CORONA-DITE $Pb_2Mn_8O_{16}$ / Imini, Antiatlas Morocco
451	black, grey, brown-black, dark brown	sub-metallic, silky opaque	$\dfrac{4}{4.3-4.4}$	perfect / uneven, brittle	columnar crystals; strongly striated, compact, granular, radial, needle-shaped / orthorhombic	in hydrothermal ore veins, sedimentary / pyrolusite, barytes, calcite & o. / pyrolusite, anti-monite	MANGANITE MnOOH (Brown Manganese Ore) / Ilfeld, Harz, Germany
452	green-grey, brownish-green, black / white, grey	vitreous, pearly, sub-metallic transluc.	$\dfrac{4}{3.2-3.3}$	very good / –	compact, granular, spathic, shelled, embedded / monoclinic	constituent in basic rocks; chemically diopside & augite with specially good cleavage / – / bronzite, hyper-sthene	DIALLAGE $Ca(Mg,Fe)$ $[Si_2O_6]$ $(Al,Fe,Ca)_2$ $[(Si,Al)_2O_6]$ (Pyroxene Family) / Diethendorf, Ore Mountains, Germany
453	black, grey-black / dark brown, black	sub-metallic, dull opaque	$\dfrac{4.5-5}{6.8-7.5}$	good / uneven, brittle	thick-tabular, short-prismatic crystals; compact, granular / monoclinic	in hydrothermal ore veins, in tin deps. / wolframite, cassiterite & o. / –	FERBERITE $FeWO_4$ Ehren-friedersdorf near Annaberg, Ore Mountains, Germany
454	brownish-black, yellowish-brown / yellowish-brown, red-brown, greenish-grey	sub-metallic, greasy transluc., opaque	$\dfrac{4-4.5}{6.4-7.1}$	very good / uneven, brittle	tabular, needle-shaped crystals; lamellar, tubular agg. / monoclinic	in tin deps. / wolframite, cassiterite, topaz, quartz & o. / –	HÜBNERITE $MnWO_4$ Slavkov, Czechoslovakia
455	black, dark brown / black, brown-black	greasy, metallic opaque	$\dfrac{5-5.5}{7.1-7.5}$	very good / uneven, brittle	thick-tabular, short-columnar, pris-matic, needle-shaped crystals; compact, radial, tubular, thin-tabular, dissemi-nated / monoclinic	in tin ore deps., in hydrothermal ore veins / cassiterite, schee-lite, molybdenite, zinnwaldite, fluorite, topaz, tourmaline	WOLFRA-MITE $(Mn,Fe)WO_4$ Cinovec, Czechoslovakia
456	black / black-green	vitreous, silky, dull opaque	$\dfrac{5}{3.6-4}$	– / brittle	compact, fine-fibrous, radial, tubular, entangled fibres, massive; palmaceous agg. / orthorhombic	metasomatic in limestone & dolo-mite, in magnetite deps. / –	LUDWIGITE $(Mg,Fe)_2Fe$ $[O_2BO_3]$ / Moravica, Banat, Romania

No.	Colour / Streak	Lustre / Transparency	Hardness / Specific Gravity	Cleavage / Fracture & o. / Phys. Props.	Common Form, Aggregates / Crystalline Syst.	Occurrence / Assoc. Minerals / Similar Minerals	Name & Chem. Formula / Origin of Specimen
457	black, brownish-black / brown	vitreous, greasy, sub-metallic opaque	5.5 / 4.5 – 4.8	— / uneven, brittle	rarely octahedral crystals; massive, granular, compact, disseminated / cubic	in ultrabasic rocks, in serpentines / olivine, bronzite & o. / magnetite, frank-linite, hercynite	CHROMITE Cr_2FeO_4 (Chromic Iron, Chrome Iron Ore) / Gulsen near Kraubath, Styria, Austria
458	black / black-brown	sub-metallic, greasy, dull opaque	5 – 6 / 4.7	ex-foliation (scaling) / con-choidal, brittle	thick-tabular, platy crystals; massive, granular, compact, rose-shaped clusters / trigonal	in basic igneous rocks, in alpine cracks, in placers / rutile, haematite, magnetite, titanite, apatite / chromite, magne-tite, haematite	ILMENITE $FeTiO_3$ (Titanoferrite) / Fleschhorn, Switzerland
459	black, red-brown, yellowish / white, grey	ada-mantine, sub-metallic, greasy opaque	5.5 / 4	good / con-choidal	cube- & octa-hedron-shaped crystals; compact, granular, reniform, pseudocubic / monoclinic	in pegmatites, basalts, crystalline schists, in meta-morphic limestones / — / —	PEROVSKITE $CaTiO_3$ / Achmatov Mine near Slatoust, Urals, USSR
460	black / black, brown	sub-metallic, greasy, dull opaque	4 – 6 / 4.4 – 4.7	— / uneven, brittle	cryptocrystalline, botryoidal, reni-form, mammilli-form, compact, stalactitic, massive, fibrous / monoclinic	in oxidized zone of manganese ore deps., hydrother-mal, sedimentary / pyrolusite, limo-nite, siderite, pyrite / pyrolusite, haus-mannite	PSILO-MELANE $(Ba,Mn)_3$ $(O,OH)_6$ Mn_8O_{16} / Příbram, Czechoslovakia
461	brown-black / whitish-brown	vitreous, sub-metallic opaque	5.5 – 6.5 / 5.1 – 6.3	— / uneven, brittle, con-choidal	columnar, bi-pyramidal crystals; granular, compact / tetragonal	in granite-pegma-tites, in placers / — / columbite, tanta-lite, gadolinite	FERGUSO-NITE $YNbO_4$ / Ytterby near Waxholm, Sweden
462	brown, black / dark yellow, brown-yellow	ada-mantine, sub-metallic, silky transluc., opaque	5 – 5.5 / 3.3 – 4.3	good / brittle	thin-lamellar, needle-shaped crys-tals; compact, granular, folia-ceous, radiating, fibrous, mammilli-form, earthy / orthorhombic	in oxidized zone of ore veins, sedi-mentary / haematite, pyrite, siderite, galena, quartz / manganite, haema-tite, lepidocrocite	GOETHITE α-FeOOH (Needle Iron-stone, Acicular Iron Ore) / Příbram, Czechoslovakia
463	yellow, brown, reddish, greenish, black / white, grey	vitreous, greasy transpar., transluc.	5 – 5.5 / 3.4 – 3.6	indistinct / con-choidal, brittle	columnar, wedge-shaped, tabular, needle-shaped crystals; compact, granular / monoclinic	in alpine cracks, in crystalline schists, in magnetite deps. / — / axinite, cf. No. 162, 339	TITANITE $CaTi[O\ SiO_4]$ (Sphene) / Arendal, Aust-Agder, Norway
464	brown, black / greenish-grey, brown	vitreous, greasy transluc., opaque	6 / 4.1	indistinct / con-choidal	thick-tabular, columnar, needle-shaped crystals; granular, compact, disseminated / monoclinic	in granites, syeni-tes, pegmatites, in volcanic ejecta / felspars, quartz & o. / gadolinite, ferguso-nite, samarskite	ALLANITE $(Ca,Ce,La,Na)_2$ $(Al,Fe,Be,Mg,$ $Mn_3)$ $[OH(SiO_4)_3]$ (Orthite) / Březová, Czechoslovakia

No.	Colour Streak	Lustre Transparency	Hardness Specific Gravity	Cleavage Fracture & o. Phys. Props.	Common Form, Aggregates Crystalline Syst.	Occurrence Assoc. Minerals Similar Minerals	Name & Chem. Formula Origin of Specimen
465	black, brown-black black ___ brown	vitreous transluc.	5 − 6 ___ 3.2	good con-choidal	short- to long-columnar crystals; no agg. monoclinic	in syenites benitoite, natrolite —	NEPTUNITE Na₂FeTi [Si₄O₁₂] San Benito County, California, USA
466	brown, dark grey, blackish-green, black ___ white, grey	vitreous opaque	5 − 6 ___ 3.5 − 3.6	clear brittle	tubular, needle-shaped crystals; granular, spathic, compact, massive monoclinic	in metamorphic & metasomatic rocks magnetite, pyrite & o. — cf. No. 343	HEDEN-BERGITE CaFe[Si₂O₆] (Pyroxene Family) Långban, Sweden
467	black, brownish-black, dark green ___ greyish-green	vitreous transluc., opaque	5 − 6 ___ 3.3 − 3.5	good con-choidal	short-columnar, thick-tabular, needle-shaped crystals; compact, granular, massive, needle-shaped, disseminated monoclinic	rock constituent in basic rocks, in tuffs, lavas & volcanic ejecta — amphibole	AUGITE (Ca,Mg,Fe₂,Fe₃, Ti,Al)₂ [(Si,Al)₂O₆] (Pyroxene Family) Lukov, Czechoslovakia
468	black, greenish-black ___ grey-white, brown	vitreous, greasy transluc., opaque	5 − 6 ___ 2.9 − 3.3	very good con-choidal	pseudohexagonal, short-columnar to acicular crystals; compact, tubular, fibrous, radiating, granular, massive monoclinic	rock constituent of many igneous rocks, in crystalline schists biotite, garnet, epidote, magnetite & o. augite, tourmaline	AMPHIBOLE (Ca,Na,K)₂₋₃ (Mg,Fe₂,Fe₃,Al)₅ [(OH,F)₂(Si,Al)₂ Si₆O₂₂] (Common & Basaltic Hornblende) Lukov, Czechoslovakia
469	black, brownish-black ___ brownish, black, brown	ada-mantine, greasy, dull transluc., opaque	6 ___ 5.2 − 6	clear uneven, brittle	thick-tabular, short-columnar crystals; compact, embedded orthorhombic	mixed crystals of niobite & tantalite, in granite-pegma-tites beryl, cryolite —	COLUMBITE (Fe,Mn) (Nb,Ta)₂O₆ (Niobite) Moss, Norway
470	black, deep blue ___ grey-green, blue-grey, whitish	vitreous transpar., transluc.	5.5 − 6 ___ 3.4 − 3.5	very good uneven, brittle	thick- & long-columnar, tabular crystals; compact, tubular, granular monoclinic	in nepheline-pegmatites, syenites nepheline, zircon, sodalite & o. aegirite, acmite	ARFVED-SONITE Na₂Fe₂(SiO₃)₄ (Amphibole Family) Kangerdlugsuak, Greenland
471	black ___ grey-black	vitreous, metallic, dull opaque	6 − 6.5 ___ 5.2 − 5.5	— brittle, uneven	pyramidal, prisma-tic, thick-tabular crystals; compact tetragonal	mixed crystals of tapiolite & rutile, in pegmatites — —	STRÜVERITE (Fe,Ti) (Nb,Ta)₂O₆ Ampangabe, Madagascar
472	black ___ brown	greasy, sub-metallic opaque	6 − 6.5 ___ 7.9 − 8	good brittle	columnar, short-acicular crystals; compact orthorhombic	in granite-pegmatites — —	TANTALITE (Fe,Mn) (Ta,Nb)₂O₁ Bodenmais, Bavarian Forest, Germany

No.	Colour / Streak	Lustre / Transparency	Hardness / Specific Gravity	Cleavage / Fracture & o. / Phys. Props.	Common Form, Aggregates / Crystalline Syst.	Occurrence / Assoc. Minerals / Similar Minerals	Name & Chem. Formula / Origin of Specimen
473	black, dark grey, brown / grey, brownish	vitreous, greasy / opaque	6 – 7 / 5.5 – 6	— / brittle	fine-crystalline, mammilliform, radiating-fibrous, crusty, massive / monoclinic	mixture of baddeleyite & decomposed zircon / — / —	Baddeleyite variety **ZIRCONIA** ZrO_2 (Zircon-favas) / Minas Gerais, Brazil
474	black, brownish-black / greenish-grey	vitreous, greasy / opaque	6.5 – 7 / 4 – 4.7	good / conchoidal, splintery	thick-tabular, short-columnar crystals; compact, massive, embedded / monoclinic	in granite-pegmatites, in alpine cracks / — / fergusonite, samarskite	**GADOLINITE** $Y_2FeBe_2[O\ SiO_4]_2$ / Ytterby near Waxholm, Sweden
475	black, blackish-green / greyish-green	vitreous, sub-metallic, dull / opaque	7.5 – 8 / 3.8	— / conchoidal	octahedral crystals; rounded grains, compact, granular / cubic	in basic igneous rocks, in metamorphic limestone / pyroxene, biotite, vesuvianite, corundum, graphite / picotite, garnet	**PLEONAST** $(Al,Fe)_2(Mg,Mn)O_4$ (Spinel Family Ceylonite) / Monzoni in Fassatal, South Tyrol, Italy
476	black / reddish-brown, dark brown	sub-metallic, metallic / opaque	6 – 7 / 4.6 – 5	poor / conchoidal, brittle	octahedral crystals; compact, granular, massive / cubic	in manganese-zinc ore deps., in metamorphic limestone / zincite, willemite, rhodonite, garnet / chromite, magnetite	**FRANK-LINITE** Fe_2ZnO_4 / Franklin, New Jersey, USA
477	deep brown, black / white	vitreous transpar., transluc.	7 / 2.6	— / conchoidal	short-columnar, pointed, hexagonal crystals; horizontally striated / trigonal	in alpine cracks, in pegmatites, in ore deps. / — / —	Quartz variety **MORION** SiO_2 / Příbram, Czechoslovakia
478	black / white, greenish	vitreous, greasy, dull / opaque	6.5 – 7 / 3.6 – 4.1	— / conchoidal, splintery, brittle	dodecahedral, icositetrahedral crystals; compact, granular, massive / cubic	in basic eruptive rocks (basalt, phonolite, nepheline-syenite) / — / pleonaste	**MELANITE** $Ca_3Fe_2[SiO_4]_3 + (Na,Ti)$ (Garnet Family) / Franklin, New Jersey, USA
479 / 480	black / white	vitreous, greasy / opaque	7 – 7.5 / 3 – 3.1	— / conchoidal to uneven, splintery	triangular, short- to long-columnar, acicular crystals; longitudinally striated; tubular, to fibrous, radiating agg.; massive, compact / trigonal	in granites, pegmatites, in crystalline schists, in metamorphic limestone & dolomite, in ore veins / muscovite, topaz, quartz, beryl, fluorite, apatite / amphibole, ilvaite	Tourmaline variety **SCHORL** $NaMg_3Al_6[(OH)_1+_3(BO_3)_3\ Si_6O_{18}]$ (Iron Tourmaline) / Siberia, USSR / **TOURMA-LINE** / Montana, USA

155

No.	Colour / Streak	Lustre / Transparency	Hardness / Specific Gravity	Cleavage / Fracture & o. / Phys. Props.	Common Form, Aggregates / Crystalline Syst.	Occurrence / Assoc. Minerals / Similar Minerals	Name & Chem. / Formula / Origin of Specimen
481	grey, yellow, brown, black / white, grey, yellowish	dull / opaque	3.5 – 4 / 3.9 – 4.2	good / conchoidal, brittle	fine-fibrous, shelled, reniform masses; crusty, striated / cubic, hexagonal	in lead-zinc ore deps., partly sphalerite, partly wurtzite / — / —	**ZINC BLENDE** ZnS / Stolberg near Aachen, Germany
482	yellow-brown, grey, yellow, black, iridescent tarnish / rusty brown, yellow	sub-vitreous, silky, dull / opaque	5 – 5.5 / 3.6 – 3.7	— / conchoidal, fibrous	cryptocrystalline, fibrous, tubular, cone-shaped, shelled, stalactitic, earthy, loose / orthorhombic	weathering product in iron ore deps. / — / cf. No. 409, 410	**LIMONITE** FeOOH . nH₂O (Brown Haematite, Brown Ironstone, Brown Iron Ore) / Železník, Czechoslovakia
483	grey to black, iridescent tarnish / black	metallic to dull / opaque	3 – 4 / 4.5 – 5	— / conchoidal, brittle	tetrahedral, dodecahedral crystals; compact, granular, massive, coated with chalcopyrite / cubic	in hydrothermal veins / galena, pyrite, bornite, chalcopyrite, siderite, quartz / cf. No. 431	**TETRA-HEDRITE** (Cu,Zn,Ag,Fe)₃ (Sb,As)S₃₋₄ (Fahl Ore) / Lizard, Cornwall, England
484	red, blue, yellow, green / white	pearly, iri-descent / transluc.	3 / 2.6 – 2.8	— / conchoidal, brittle	massive limestone containing fossil mollusc shells / —	in sedimentary limestones / — / —	**LUMACHEL** CaCO₃ (Shell Marble) / Bleiberg, Carinthia, Austria
485	yellow, white, red-brown, violet, grey, brown / white, yellowish	dull, silky / opaque	3.5 – 4 / 2.9 – 3	— / fibrous separation, brittle	fibrous agg.; concentrically shelled, stalactitic, crusts, sinter formations / orthorhombic	aragonite sinter, deposited at hot springs / — / cf. No. 320	**SPRUDEL-STEIN** CaCO₃ (Aragonite) / Carlsbad, Czechoslovakia
486	white, yellow, grey, brown, red / white	dull / transluc., opaque	5.5 – 6 / 2.1 – 2.2	— / porous, rough	crusty, botryoidal, pea-shaped, earthy, loose, coatings / amorphous	opal sinter, deposited at geysers & hot springs / — / —	**GEYSERITE** SiO₂ . nH₂O (Siliceous Sinter, Opal) / Stóra Geysir, Iceland
487	white, yellow, brown, grey / white	greasy / transluc., opaque	3.5 – 4 / 5.8 – 6.4	— / conchoidal	radiating, needle-shaped, globular agg.; cryptocrystalline / hexagonal	calcareous pyromorphite, in lead-zinc ore deps. / galena, sphalerite, pyrite, barytes / —	**POLY-SPHAERITE** (Pb,Ca)₅ [Cl(PO₄)₃] (Miesite) / Stříbro, Czechoslovakia
488	blue-green, yellowish, brownish, greenish, gold-brown / blue-grey, gold-brown, yellow	silky; dull / transluc., opaque	5.5 – 6 / 3.2 – 3.3	very good / fibrous	parallely fine-fibrous agg.; zonally contorted, cracked, platy / monoclinic	fibrous riebeckite asbestos, as crack filling in fine-grained quartz rocks, in crystalline schists, partly silicified grey-blue falcon eye, partly decomposed & yellow tiger eye / — / cf. No. 543	**CROCI-DOLITE** Na₂Fe₃Fe₂ [(OH,F) Si₈O₁₁]₂ (Amphibole Family, Riebeckite Asbestos, Tiger's Eye, Cat's Eye) / Griquatown, South Africa

No.	Colour Streak	Lustre Transparency	Hardness Specific Gravity	Cleavage Fracture & o. Phys. Props.	Common Form, Aggregates Crystalline Syst.	Occurrence Assoc. Minerals Similar Minerals	Name & Chem. Formula Origin of Specimen
489	green, yellow, black, red, brown, spotted __white__, grey	greasy, dull transluc., opaque	$3-4$ __$2.5-2.6$__ conchoidal, soft to splintery	—	microcrystalline, foliaceous, scaly, massive, compact, disseminated __monoclinic__	fillings in veins & cracks in peridotites & grained limestone; in pseudomorphs after olivine __chromite, magnetite, opal, talc & o.__ —	**SERPENTINE** $Mg_6(OH)_8$ $[Si_4O_{10}]$ (Antigorite) __Lizard, Cornwall, England__
490	grey-blue, grey-yellow, violet, bluish, greenish __white, grey__	silky transluc., opaque	$3-4$ __$2.3-2.5$__	good flexible to brittle	fine-fibrous, capillary crystals; felt-like, parallelly fibrous agg. __monoclinic__	filling in veins & cracks in serpentines __olivine, tremolite & o.__ — cf. No. 255	**ASBESTOS** Mg_6 $[(OH)_8Si_4O_{10}]$ (Chrysotile) __Newjansk near Sverdlovsk, Urals, USSR__
491	yellow, brown, black, grey __white, grey__	vitreous, dull __opaque__	7 __$3.1-3.2$__	uneven, splintery, brittle	long-columnar, rounded crystals; with cruciform dark core (pigmented by some carbonaceous substance) __orthorhombic__	in metamorphic rocks, in clay schists __staurolite, cordierite, tourmaline__ —	Andalusite variety **CHIASTOLITE** $Al_2[O SiO_4]$ (Macle) __Lancaster, Lancashire, England__
492	colourless, blue, yellow, greenish, grey __white__	vitreous, pearly transpar., transluc.	4 along cleavage planes, 6 across cleavage planes __$3.6-3.7$__	very good __brittle__	wide-tubular, lamellar crystals; fibrous, radiating, foliaceous __triclinic__	in crystalline schists, pegmatites, granulites __staurolite, andalusite, corundum__ sillimanite, cf. No. 236	**CYANITE** $Al_2[O SiO_4]$ (Disthene) __Pizzo Forno, Tessin, Switzerland__
493 494	colourless, green, red, brown, yellow, black __white__	vitreous transpar., transluc.	$7-7.5$ __$2.9-3.2$__	uneven, splintery	triangular, columnar crystals; longitudinally striated, zonal colouring __trigonal__	in pegmatites, in granites & metamorphic rocks — — cf. No. 208, 302, 479, 480	**TOURMALINE** $NaFe_3Al_6[(OH)_4$ $(BO_3)_3Si_6O_{18}]$ __Minas Gerais, Brazil__ **TOURMALINE** $NaFe_3Al_6[(OH)_4$ $(BO_3)_3Si_6O_{18}]$ __Grotta d'Oggi, Elba__
495 496	white, grey, green, blue, red, yellow, iridescent __white__	pearly, greasy, opalescent transluc.	$5.5-6.5$ __$1.9-2.5$__	conchoidal to uneven, brittle, strong internal reflections	belts & veins, irregular nodules, compact, disseminated, concretions __amorphous__	in cavities in young volcanic igneous rocks, in sandstones, conglomerates, clays; due to decomposition of siliceous rocks — cf. No. 535	Opal variety **PRECIOUS OPAL** $SiO_2 . n H_2O$ __Cooper Creek, Queensland, Australia__ Opal variety **PRECIOUS OPAL** $SiO_2 . n H_2O$ __Bulls Creek, Queensland, Australia__

159

No.	Colour / Streak	Lustre / Transparency	Hardness / Specific Gravity	Cleavage / Fracture & o. / Phys. Props.	Common Form, Aggregates / Crystalline Syst.	Occurrence / Assoc. Minerals / Similar Minerals	Name & Chem. Formula / Origin of Specimen
497	white, grey, green, blue, red, yellow, iridescent / white	pearly, greasy, opalescent / transluc.	5.9 – 6.5 / 1.9 – 2.5	— / conchoidal to uneven, brittle	bands & veins, irregular nodules, compact, disseminated, concretions / amorphous	— / — / — / cf. No. 495, 535	Opal variety **PRECIOUS OPAL** $SiO_2 . n\ H_2O$ / Bulls Creek, Queensland, Australia
498	white, grey, black / white	vitreous, greasy / transluc.	6 – 7 / 1.9 – 2.5	— / conchoidal, brittle	irregular nodules, bands, veins, massive / amorphous	milk opal with embedded black, mossy, arborescent structures composed of foreign substances / —	**MOSSY OPAL** $SiO_2 . nH_2O$ / Kansas near Collyer, Trego Co., USA
499	red, yellow, red-brown, brown / yellowish	vitreous, dull / opaque	7 / 2.5 – 2.6	— / conchoidal, brittle, splintery	nodules, grains, concretions, massive, crusty / trigonal	quartz containing iron oxides & iron hydroxides, in iron ore deps. / — / —	Quartz variety **EISEN-KIESEL** $SiO_2 + (Fe_2O_3)$ / Zaječov, Czechoslovakia
500	yellow-brown, red, white, yellowish / white	vitreous with metallic lustre / opaque	6 – 6.5 / 2.6 – 2.7	very good / uneven, brittle	thick-tabular, massive, granular, spathic / triclinic	plagioclase containing mica or haematite scales, constituent in granites, syenites / granites, syenites	**SUNSTONE** $Ca[Al_2Si_2O_8]$ $Na[AlSi_3O_8]$ (Plagioclase Family, Aventurine Felspar) / Frederiksvärn, Norway
501	grey, white, yellow, brown, bluish / white	vitreous, dull / transluc.	7 / 2.6	— / conchoidal	fine-fibrous, reniform, botryoidal, mammilliform, crusts, nodules, concretions / trigonal	micro-crypto-crystalline quartz, in cracks & cavities of basaltic rocks / — / cf. No. 422	**CHALCEDONY** SiO_2 / Tampa Bay, Florida, USA
502	red, white / white	vitreous transluc., opaque	7 / 2.5 – 2.6	— / conchoidal, brittle	massive, compact, nodules, pebbles, stratified, crypto-crystalline / trigonal	white-red stratified chalcedony / — / —	Chalcedony variety **CORNELIAN ONYX** SiO_2 / Brazil
503	yellow-green, brown, grey, variously coloured / white	greasy, dull / opaque	7 / 2.5 – 2.6	— / uneven, splintery	finely granular, massive, compact, nodules, irregular masses / trigonal, amorphous	chalcedony with admixtures, in cracks, fissures & cavities in different rocks / —	**JASPER** SiO_2 / Cairo, Egypt
504	multi-coloured / white	vitreous, greasy, dull / transluc., opaque	7 / 2.5 – 2.6	— / conchoidal	finely granular, massive, compact, stratified, nodules, pebbles, irregular masses / trigonal, amorphous	variegated chalcedony composed of different coloured bands, in vesicles & cavities, in amygdaloidal volcanic rocks / — / —	**AGATE** SiO_2 / India

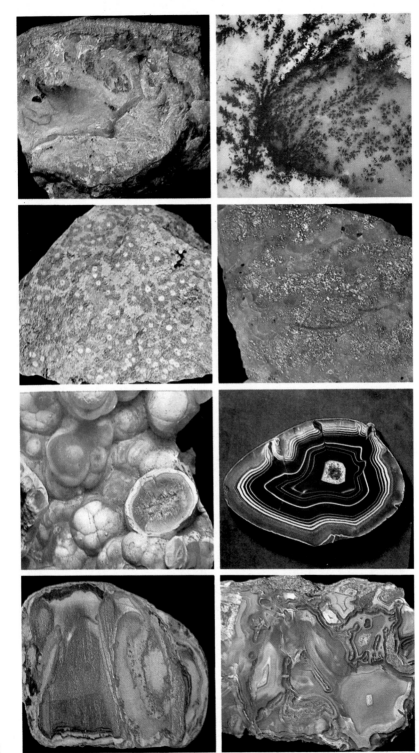

No.	Colour / Streak	Lustre / Transparency	Hardness / Specific Gravity	Cleavage / Fracture & o. Phys. Props.	Common Form, Aggregates / Crystalline Syst.	Occurrence / Assoc. Minerals / Similar Minerals	Name & Chem. Formula / Origin of Specimen
505	red, yellow, greenish, variegated / white	vitreous, greasy, dull / transluc., opaque	7 / 2.5 − 2.6	— / conchoidal	most finely granular, massive, nodules, pebbles, compact, striated / trigonal	stratified, in cross fracture of banded chalcedony, partly with parallel bands, as filling in amygdaloidal volcanic rocks / — / —	AGATE SiO_2 / St. Wendel, Saar, Germany
506	black, white / white	vitreous / transluc., opaque	7 / 2.5 − 2.6	— / conchoidal	massive, nodules, amygdale filling, pebbles, irregular masses, crypto-crystalline / trigonal	white-black-banded chalcedony / — / —	Chalcedony variety ONYX SiO_2 / Minas Gerais, Brazil
507	grey, white, red, brown, variegated / white	vitreous, greasy, dull / transluc., opaque	7 / 2.5 − 2.6	— / conchoidal	most finely granular, massive, nodules, irregular masses, pebbles, finely shelled structure / trigonal, amorphous	filling in cavities & amygdales in basalts / — / —	AGATE SiO_2 / Levín, Czechoslovakia
508	grey to black, white / white	vitreous / transluc., opaque	7 / 2.5 − 2.6	— / conchoidal	massive, compact, pebbles, amygdaloidal, crypto-crystalline / trigonal	white-grey-banded chalcedony, filling in amygdaloidal volcanic rocks / — / —	Chalcedony variety ONYX SiO_2 / Uruguay
509							AGATE SiO_2 / India
510	variegated / white	vitreous, greasy, dull / transluc., opaque	7 / 2.5 − 2.6	— / conchoidal	most finely granular, massive, irregular masses, rounded debris, nodules, layered structure, stratified / trigonal	filling in cavities, cracks & vesicles in extruded rocks, especially basalts / — / —	AGATE SiO_2 / Měděnec, Czechoslovakia
511							AGATE SiO_2 / Ratnapura, Ceylon
512	blue, yellow, colourless, red, brown, grey / white	vitreous, greasy / transpar., transluc.	9 / 3.9 − 4.1	separation planes / conchoidal, splintery	short-columnar, fusiform, thick-tabular crystals; granular, compact, massive / trigonal	in pegmatites, crystalline schists, in metamorphic limestones & dolomites, in granite, syenite & basalt, in placers / — / zircon, spinel, garnet, cordierite, apatite, topaz, tourmaline, cf. No. 428, 514, 515	CORUNDUM Al_2O_3 / Ceylon

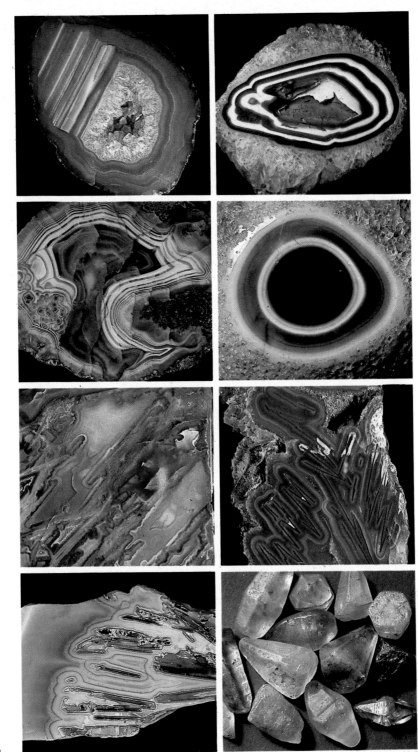

Precious Stones

The large family of precious stones includes all minerals of beautiful appearance which catch our eye with their sparkling colours, transparency and lustre. A precious stone must be to a certain extent indestructible and possess the capacity of resisting wear. Only the harder minerals (7 – 10) possess such properties, preserving their beautiful appearance even when subjected to being worn as jewellery. The most valuable precious stones are those which are rarely found naturally. Fashion — unpredictable as it is — sometimes plays an important part in determining the popularity of the various kinds.

Precious stones do not form any natural group of minerals, their number varying with the whim of fashion. The most enduring favourites — unaffected by fashion — are the diamond, the ruby, the sapphire and the emerald. These four stones occupy a special position, and are considered the 'true precious stones', others of lesser value being called 'semi-precious'. As it is not easy to draw a definite boundary between precious and semi-precious stones, it is preferable to use the term 'gemstones' for the latter.

Since the primeval age of history precious stones have always been associated with man. Primitive people appreciated their value as jewellery, and knew perfectly well how to make full use of their special qualities. Precious stones were believed to possess some mysterious, magical properties, and a special healing power. Stones were thought to be connected with the stars and were believed to be able to influence human fate. Therefore everybody should carry his so-called 'birth-stone' with him (the stone for the month in which he was born), as this was supposed to safeguard his health, happiness and property. Independent of the stars, however, every stone worn as an amulet was thought to possess a special power, as Goethe wrote in his charming verses:

> 'If carnelian your mascot be
> brings happiness and welfare thee;
> It drives away all wrong apace
> protecting you and all your place.'

As with many other superstitions, this one has survived until the present day, and every modern jeweller will recommend the right 'birth-stone' for you. In the present technical world the importance of precious stones — which are used in a number of different industries — has been growing continuously. Last but not least let us mention space research which would be impossible without precious stones.

Treatment of Precious Stones — Cutting Form

At first precious stones were used in their natural form. Only later people began to polish them to stress their colour and lustre. The art of treating precious stones was best understood by the Indians and Egyptians. The optical properties of transparent stones enhanced by a symmetrical cut of small, smooth surfaces called facets.

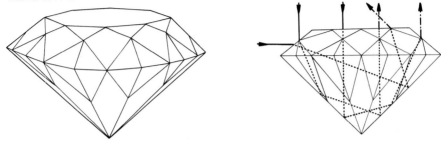

Brilliant Cut and Light Refraction in a Cut Stone

When cutting precious stones several different cuts may be applied, from the simplest step cut to the most complicated cuts with many facets. The so-called brilliant cut has been popular since the 17th century. It is used for diamonds as well as for other transparent precious stones. In this cut a special arrangement of facets causes a repeated reflection within the mineral and intensifies the rays of light emerging from it. In this way the lustre and colour of the mineral are enhanced and a special brilliance is achieved. Light refraction causes the splitting-up of each ray of white light into its coloured constituents. The result is a 'play of colours' better known as a 'play of fire'.

Opaque, coloured minerals are often cut into facetless domed forms. This treatment is called the Cabochon Cut. The special art and science of gem-engraving is called glyptography, at which the ancient Greeks and Romans excelled.

When treating a precious stone the expert must be perfectly acquainted with his material. Moreover he must know how to make full use of its physical properties, i.e. the cleavage, hardness and pleochroism of the individual mineral types. When mounting a stone in a jewel the jeweller must also take into account its power of resisting high temperature, pressure, and acids as well as its shock resistance.

The unit of weight for gemstones is a carat which equals 0.2 gm; 1 gm = 5 carats.

Imitations and Synthetic Precious Stones

Because of their rarity and consequent high value precious stones have always tempted people to try to imitate them. As early as in the Middle Ages imitation precious stones were being made of glass, others were artificially coloured or lined underneath with silver foil to achieve a brighter lustre. A special sort of imitation is the so-called doublet. It consists of two stones cemented together. Sometimes genuine stones are cemented together with glass, or two smaller genuine pieces may be cemented to each other. All imitations and counterfeits merely have the appearance of genuine stones but not their physical properties.

As with gold, alchemists tried in vain to find some nostrum that would help them to change inferior stones into gemstones. The artificial production of precious stones is the achievement of modern 'alchemists'. Stones produced in laboratories are equal to natural stones, i.e. they have the same chemical as well as physical properties. Consequently, synthetic gems are genuine stones but not natural products. One of the oldest and most successful techniques for artificial production of gems is the Verneuil process. In this process gems are made by remelting fine powdery raw material in an oxy-hydrogen flame. Large numbers of rubies, sapphires, spinels, rutiles and others that do not occur naturally, are still obtained in this way. Stones (called boules) made by the Verneuil process are of cylindrical form bounded by no straight plane.

Synthetic quartz (rock crystal, smoky quartz and citrine), is produced under high pressure and temperature in steel bombs by so-called hydrothermal synthesis. Synthetic emeralds can be produced from meltings under high temperatures in the form of perfect crystals.

The diamond — the king of precious stones — is now produced artificially under high pressure and temperature in special furnaces. The crystals obtained are too small to be used for jewellery and like most of the synthetic precious stones, they are used extensively in industry.

Identification of Precious Stones

Even the most experienced specialist cannot positively identify a precious stone at first sight, especially since there are so many perfect imitations. It is usually not difficult to tell a genuine precious stone from a glass imitation. One can thus distinguish an aquamarine from its synthetic imitation, the spinel, since both these stones differ in their chemical and optical properties.

A number of precious stones can be identified by means of the enclosed tables. Others cannot be identified so easily and require accurate measurements of the important optical properties such as the refraction index, double refraction, pleochroism and so on, which require special apparatus.

How can one distinguish synthetic from genuine precious stones? This question is often difficult to answer since they differ practically only in the manner of their growth. A powerful microscope, however, reveals different inclusions in the stones, minute bubbles and other characteristic internal markings. In genuine stones such impurities are usually spaced along the crystal planes whereas in stones produced by the Verneuil process curved growth lines may be observed, and the inclusions are lenticular forming flat-topped shells. In some synthetic stones, such as emerald, obtained by certain techniques, even the microscopic examination does not provide a reliable proof.

In this book precious stones are arranged according to their value and according to natural groups commonly used in the study of gemstones. The enclosed tables provide examples of synthetic stones completed with refractive index and references to the relevant numbers of the same minerals in the preceding mineral tables.

Apart from the popular minerals already mentioned, moldavite is quoted amongst the precious stones. It belongs to the so-called tektites. It is a refractory glass with a schlieric (streaked) texture. The moldavites have a rather irregular form, are dark-green or bottle-green, and their surfaces have a pitted or grooved appearance.

Tektites are found in only a few places in the world. Apart from meteorites they have been considered, and are still sometimes believed, to be of extraterrestrial origin. According to recent theory they are more likely to originate from terrestrial sediments which were remelted when large meteorites crashed to the earth's surface some time ago. Their origin, however, has not yet been satisfactorily explained.

In the past moldavites were cut as jewels, and imitations were often made from green bottle glass. At the moment it is fashionable to use moldavites as jewels in their original rough form.

Rocks

Rocks were mentioned only briefly in the introductory chapter. Individual rocks are composed of minerals each of a certain definite composition. This composition depends upon the kind of material and on the conditions prevailing at the time of their origin. According to their structure rocks may be divided into the following three large classes:

1. magmatic rocks (igneous or eruptive rocks)
2. sedimentary rocks (deposited rocks)
3. metamorphic rocks (alteration rocks, crystalline schists)

Classification of Rocks

Magmatic rocks may be divided according to the position in which the hot molten magma had solidified.

Intrusive, or plutonic rocks (called after Pluto, the God of the Underworld). The magma crystallized below the surface of the earth.

Extrusive, or volcanic rocks (after Vulcan, the God of Fire). In a volcanic eruption some of the liquid magma poured out upon the earth's surface and solidified there.

Veinstones. The magma invaded and penetrated fissures and cracks in the adjacent rocks and crystallized there.

According to their composition, acidic rocks (rich in silicates, SiO_2) are distinguished from basic, or alkaline rocks (low in SiO_2). Acidic rocks are mostly light in colour, and are mainly composed of quartz, felspars (orthoclase, plagioclase, microcline), felspar derivatives (nepheline, leucite, sodalite), and muscovite. Alkaline rocks, on the other hand, are composed of dark minerals, such as amphibole, pyroxene, biotite and olivine.

The most important magmatic rocks can be divided according to their diminishing content of silica. The main constituents are indicated in brackets.

Intrusive rocks: granite (quartz, felspar, mica, amphibole, pyroxene), syenite (felspar, mica, amphibole, pyroxene), diorite (felspar, amphibole, pyroxene, biotite), gabbro (felspar, amphibole, augite, biotite, olivine), peridotite (olivine, pyroxene, amphibole).

Extrusive rocks: quartz porphyry and liparite (quartz, felspar, biotite, pyroxene, amphibole), porphyry and trachyte (felspar, biotite, amphibole, pyroxene), phonolite (felspar, pyroxene, nepheline or leucite), andesite (felspar, biotite, amphibole, pyroxene), basalt and melaphyre (felspar or nepheline, leucite, pyroxene, olivine, glass), dolerite (felspar, pyroxene, partly olivine), glass.

Veinstones: aplites and pegmatites (their composition resembles that of intrusive rocks). Aplites are finely granular, pegmatites in extreme cases are coarsely granular.

Sedimentary rocks or deposited rocks are secondary rocks which are derived from decomposed and weathered older rocks. Therefore they are the products of decomposition which have been transported by water, ice or wind to some other place and deposited there. According to the way in which they were formed we distinguish the following kinds of sediments:

1. clastic sediments
2. chemical sediments
3. organic (biogenic) sediments.

Clastic sediments (weathered particles separated according to their specific weight and grain size) can be loose, as in gravel (broken stone), loess, and sand. In conglomerates or breccias, sandstones and claystones, the loose particles are cemented by clay, lime or quartz.

Chemical sediments originate from substances dissolved in water which have precipitated, either due to evaporation or to the change in the chemical composition of the solution. They include rock salt, gypsum, limestone, limonite, and bauxite.

Organic sediments, such as chalk, some calcareous rocks and coal, originated as a result of accumulations of the remains of dead animals and plants.

In sedimentary rocks, apart from the manner of their origin and their composition, the size of their grains is also important. Coarse and angular pieces cemented by a binding agent are called breccias. Coarsely cemented rounded pebbles are called conglomerates. In sandstones the grains are much smaller and sometimes cannot be distinguished by the naked eye. Similar to sandstones are the greywackes (quartz, felspar, and other igneous fragments) and the arkoses (quartz and felspar). Grains smaller than 0.02 mm are found in clays (kaolinite and montmorillonite) and limestones such as marl, chalk, calcite and dolomite, and also in quartz.

The *metamorphic rocks* originated from sedimentary or magmatic rocks under the influence of high pressures and temperatures deep within the earth. The original constituents of these rocks were remelted, mixed, compressed or folded, and crystallized to form new minerals. Metamorphic rocks are similar in composition to magmatic rocks, but their constituents are usually arranged in stratified layers similar to those of sedimentary rocks.

The alteration of acidic magmatic rocks results in the formation of gneiss schist and granulite; greenstone, amphibole and serpentinite come from alkaline rocks. Through recrystallization marble originates from limestones, and quartzite from sandstones. A slaty mixture of quartz and mica produces mica schist, claystone or marl produces phyllite.

Apart from the rock forming minerals already mentioned under magmatic rocks, the following minerals are especially characteristic of the alteration rocks: kyanite, sillimanite, zoisite, straurolite, garnet, talc, etc.

Structure and Texture of Rocks

The structure of rocks, i.e. the spatial distribution of minerals in rocks is an important external characteristic feature enabling them to be classified. There is a difference between the structure (manner in which rocks are formed) and the texture (arrangement of constituent parts). Structure means the mutual relationship of individual constituents which is derived from the manner of origin of the rock. According to the size of the constituents a coarse or fine-crystalline to glassy structure will result. If larger crystals are disseminated in the uniform rock matrix, the structure is called porphyritic.

Texture means the spatial distribution of individual constituents. In magmatic rocks the texture can be banded, uniform, massive, porous, or amygdaloidal. Altered (metamorphic) and sedimentary rocks have a coarse- to fine-layered, foliaceous, fibrous, or striated texture.

Identification of Rocks

With our simple techniques only an approximate identification of rocks can be made. More exact identification being possible only with the aid of a polarising microscope.

Texture, structure and jointing are important and characteristic properties of rocks. They often enable a determination of the rock's origin to be made. By means of a pocket lens individual mineral constituents may be identified. Note carefully the proportion of the main constituents, such as quartz, felspars, mica, amphibole, pyroxene, olivine, and carbonates. The identification is made easier when the minerals forming the rocks are large.

Precious Stones and Rocks

Identification Tables and Plates

PRECIOUS STONES AND GEMSTONES

No.	Colour / Streak	Lustre / Transparency / Refractive Index	Hardness / Specific Gravity	Cleavage / Fracture & o. / Phys. Props.	Common Form, Aggregates / Crystalline Syst.	Occurrence / Assoc. Minerals / Similar Minerals	Name & Chem. Formula / Origin of Specimen
513	colourless, yellow, blue, red, green, grey / white	adamantine / transpar. / 2.42	10 / 3.5	perfect / conchoidal / brittle	octahedral, cube-shaped, dodecahedral crystals; often bulbous & striated faces / cubic	embedded in ultra-basic rocks, in conglomerates, in sands, in placers / — / cf. No. 118	DIAMOND / C / Kimberley, South Africa
514	blue, greyish, white / white	vitreous / transluc. / 1.76	9 / 3.9 − 4	separation planes / conchoidal, splintery	rounded crystals & grains / trigonal	corundum with stellate inclusions, sapphire with stellate opalescence; in placers of precious stones, in channel beds / — / cf. No. 428, 512	Corundum variety ASTERIATED SAPPHIRE / Al_2O_3 / Sabaragamuwa, Ceylon
515	colourless, red, blue, violet, yellow, orange / white	vitreous / transpar. / 1.76	9 / 3.9 − 4	— / conchoidal, splintery	cylindrical, rounded 'grains' / trigonal	synthetic corundum produced by Verneuil process / — / cf. No. 428, 512	CORUNDUM / Al_2O_3 / — / synthetic
516	green / white	vitreous / transpar., transluc. / 1.57	7.5 − 8 / 2.6 − 2.8	— / conchoidal, brittle	hexagonal, short-columnar, thick-tabular crystals / hexagonal	produced synthetically / — / cf. No. 306	EMERALD / $Al_2Be_3[Si_6O_{18}]$ / (Beryl) / synthetic
517	pale green, pale blue / white	vitreous / transpar. / 1.57	7.5 − 8 / 2.6 − 2.8	— / conchoidal, brittle	short- to long-columnar crystals; tubular agg.; loose grains, pebbles / hexagonal	cf. No. 305	Beryl variety AQUAMARINE / $Al_2Be_3[Si_6O_{18}]$ / Itambacuri, Brazil
518	yellow, yellowish-green / white	vitreous / transpar., opalescent / 1.57	7.5 − 8 / 2.6 − 2.8	— / conchoidal, brittle	short- to long-columnar crystals; tubular agg.; loose grains, pebbles / hexagonal	cf. No. 305	HELIODOR / $Al_2Be_3[Si_6O_{18}]$ / (Beryl) / Nertschinsk, Transbaikal, USSR
519	rose / white	vitreous / transpar., transluc. / 1.58	7.5 − 8 / 2.6 − 2.8	— / conchoidal, brittle	short-columnar, thick-tabular crystals / hexagonal	cf. No. 305	MORGANITE / $Al_2Be_3[Si_6O_{18}]$ / (Rose Beryl) / Pala Chief, California, USA
520	green, red, blue, brown, black, multicoloured / white	vitreous / transpar., transluc. / 1.62 to 1.64	7 − 7.5 / 3 − 3.1	— / uneven, brittle, splintery	triangular, short- to long-columnar crystals / trigonal	in pegmatites, granites, metamorphic rocks / — / cf. No. 302, 479, 493, 494	TOURMALINE / $NaFe_3Al_6$ $[(OH)_4(BO_3)_3Si_6O_{18}]$ / Grotta d'Oggi, Elba, Italy

514

516

518

520

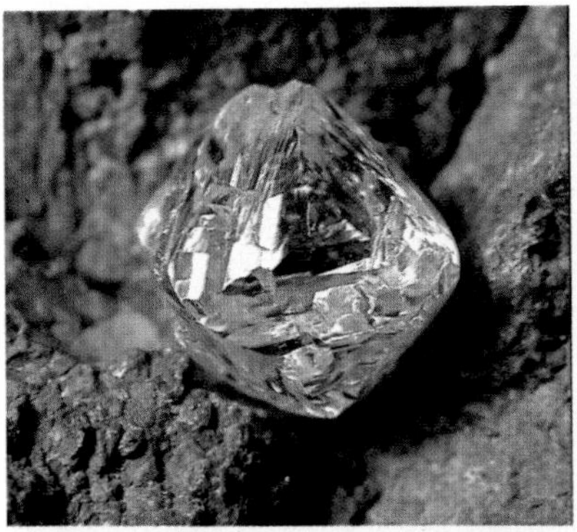

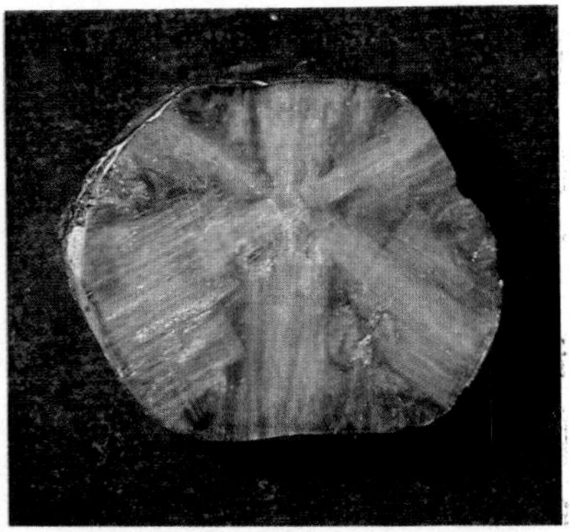

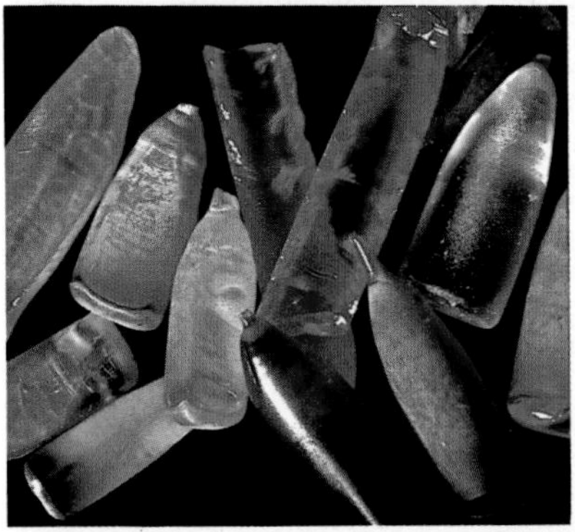

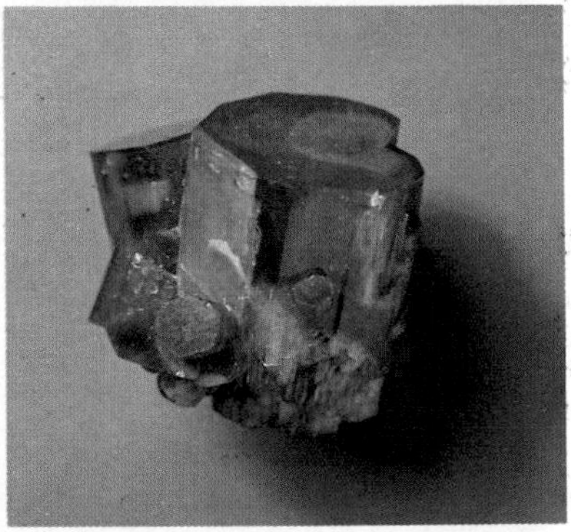

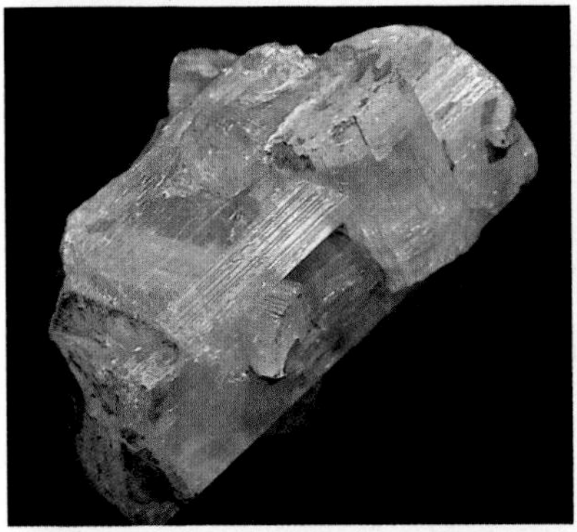

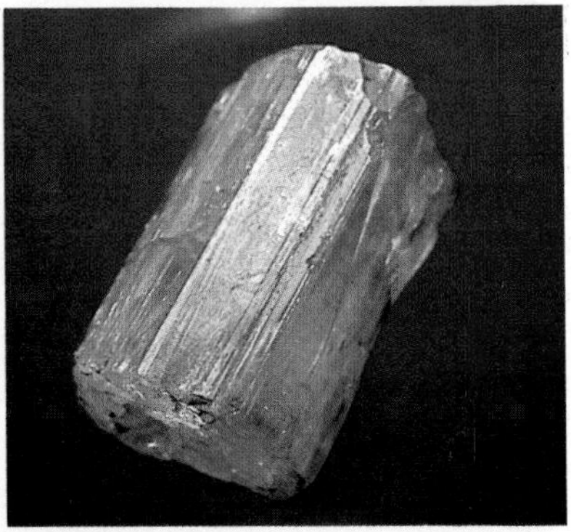

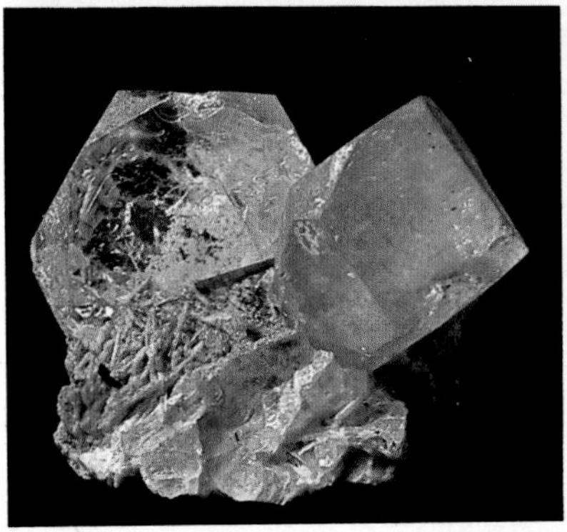

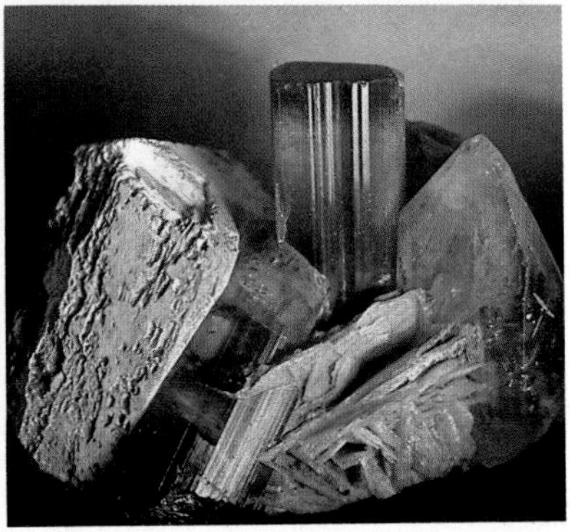

PRECIOUS STONES AND GEMSTONES

No.	Colour Streak	Lustre Transparency Refractive Index	Hardness Specific Gravity	Cleavage Fracture & o. Phys. Props.	Common Form, Aggregates Crystalline Syst.	Occurrence Assoc. Minerals Similar Minerals	Name & Chem. Formula Origin of Specimen
521	colourless, blue, red, orange, yellow, green, brown / white	vitreous / transpar., transluc. / 1.92 to 1.98	7—7.5 / 4.2—4.7	— / conchoidal, brittle	short-columnar to acicular crystals; rounded, loose grains / tetragonal	in volcanic rocks, in crystalline schists, in placers of precious stones / — / garnet, rutile, vesuvianite, sapphire, spinel, cf. No. 355	ZIRCON $Zr[SiO_4]$ / Ceylon
522	blue & in all colours / white	vitreous / transpar. / 1.72	8 / 3.5—3.7	poor / conchoidal, brittle	cylindrical, rounded 'grains' / cubic	produced synthetically by Verneuil process / — / cf. No. 246, 360	SPINEL Al_2MgO_4 / synthetic
523	colourless, yellow, blue, red, rose, brown / white	vitreous / transpar. / 1.61	8 / 3.5—3.6	perfect / uneven	short- to long-columnar crystals; with orthorhombic outline, compact, granular, massive / orthorhombic	in granite-pegmatites, in tin ore veins, in placers of precious stones / — / quartz, corundum, phenakite, beryl, aragonite, cf. No. 117	TOPAZ $Al_2[F_2SiO_4]$ / Schneckenstein near Falkenstein, Ore Mountains, Germany
524	green in daylight, red in artificial light / white	vitreous / transpar., transluc. / 1.75	8.5 / 3.6—3.7	good / conchoidal	tabular, pseudo-hexagonal triplets, striated / orthorhombic	in granite-pegmatites, granites, gneiss, mica schists / — / olivine, beryl, apatite, tourmaline, corundum, cf. No. 304	ALEXANDRITE Al_2BeO_4 (Chrysoberyl) / Takovaja River, Central Urals, USSR
525	green, greenish-yellow / white	vitreous, greasy / transpar. / 1.8 to 1.9	6.5—7.5 / 3.8—3.9	poor / conchoidal, splintery	dodecahedral, grown-up crystals / cubic	in cracks & crystalline schists, in gold placers / — / cf. No. 300	Andradite variety DEMANTOID $Ca_3Fe_2[SiO_4]_3$ (Garnet Family) / Ala, Piedmont, Italy
526	red, violet / white	vitreous, greasy / transpar. / 1.7 to 1.8	7 / 4.1—4.3	uneven, brittle, splintery	rhombdodecahedral crystals; rounded grains / cubic	in gneiss & mica schists, in placers of precious stones / — / cf. No. 219	ALMANDINE $Fe_3Al_2[SiO_4]_3$ (Garnet Family) / India
527	blood-red / white	vitreous / transpar. / 1.7 to 1.8	7—7.5 / 3.7—3.8	— / conchoidal, splintery	rounded grains without crystalline boundary, embedded or weathered, loose grains / cubic	in serpentine, peridotite, kimberlite, secondary in sands, conglomerates, in placers / —	PYROPE $Mg_3Al_2[SiO_4]_6$ (Garnet Family, Bohemian Garnet) / Podsedice, Czechoslovakia
528	yellow-green / white	vitreous / transpar. / 1.6 to 1.7	6.5—7 / 3.3—3.5	good / conchoidal	thick-tabular, short-columnar crystals; granular to massive, granular nodules, insets / orthorhombic	in basic volcanic rocks, in magnetite deps. / — / chrysoberyl, cf. No. 299	OLIVINE $(Mg,Fe)_2[SiO_4]$ (Chrysolite, Peridot) / Podmoklice, Czechoslovakia

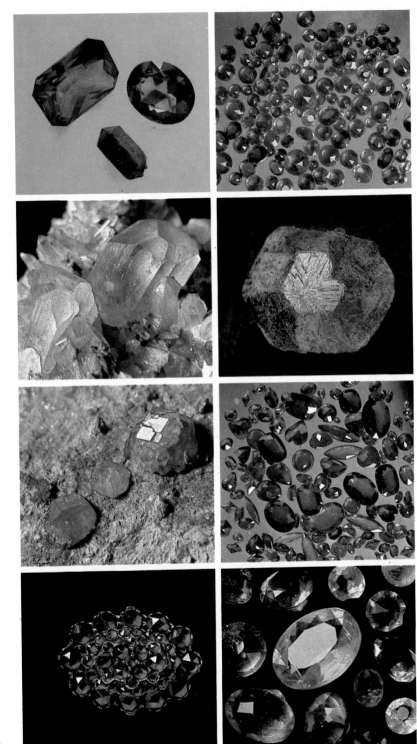

No.	Colour / Streak	Lustre / Transparency / Refractive Index	Hardness / Specific Gravity	Cleavage / Fracture & o. / Phys. Props.	Common Form, Aggregates / Crystalline Syst.	Occurrence / Assoc. Minerals / Similar Minerals	Name & Chem. Formula / Origin of Specimen
529	pale green, yellow / white	vitreous / transpar. / 1.67	6.5−7 / 3.1−3.2	good / conchoidal	columnar, thick-tabular crystals; compact, wide-tubular, spathic / monoclinic	in granite-pegmatites & crystalline schists / — / scapolite, cf. No. 108	Spodumene variety HIDDENITE LiAl[Si$_2$O$_6$] / Figueira, Brazil
530	violet, rose / white	vitreous / transpar. / 1.67	6.5−7 / 3.1−3.2	good / conchoidal	columnar, thick-tabular, longitudinally striated crystals; spathic, compact, wide-tubular / monoclinic	in granite-pegmatites / lepidolite, rubellite amblygonite, scapolite, amethyst, cf. No. 108	Spodumene variety KUNZITE LiAl[Si$_2$O$_6$] / Pala Chief, California, USA
531	colourless / white	vitreous / transpar. / 1.55	7 / 2.65	— / conchoidal	columnar to acicular crystals in druses / trigonal	hydrothermal formation, in cracks & fissures, in magmatic rocks, crystalline schists, in ore veins / — / phenakite, topaz, beryl, cf. No. 109	QUARTZ SiO$_2$ (Rock Crystal) / Hot Springs, Arkansas, USA
532	light yellow, gold-yellow / white	vitreous / transpar. / 1.55	7 / 2.65	— / conchoidal	columnar crystals; in druses, loose rounded grains in placers / trigonal	cf. No. 167	Quartz variety CITRINE SiO$_2$ / Rio Grande do Sul, Brazil
533	violet / white	vitreous / transpar. / 1.55	7 / 2.65	— / conchoidal	columnar crystals in druses / trigonal	in druses, in volcanic extruded rocks / — / apatite, cf. No. 217	Quartz colour variety AMETHYST SiO$_2$ / Capnic, Romania
534	clove-brown, smoky grey / white	vitreous / transpar. / 1.55	7 / 2.65	— / conchoidal	columnar crystals in druses / trigonal	in pegmatites, in alpine cracks, in ore deps. / — / brown topaz	Quartz colour variety SMOKY QUARTZ SiO$_2$ / Gotthard area, Switzerland
535	white, grey, green, blue, red, yellow, brilliant play of colours / white	pearly, greasy / transluc. / 1.45	5.5−6.5 / 1.9−2.5	— / uneven, brittle, strong internal reflexion	bands & veins, irregular nodules, compact, disseminated, concretions / amorphous	in cavities in young volcanic rocks / — / cf. No. 495	PRECIOUS OPAL SiO$_2$. n H$_2$O / Dubník, Czechoslovakia
536	light rose, rose-red / white	greasy / transluc. / 1.55	7 / 2.65	— / conchoidal, brittle, striated	only compact, disseminated, massive, large-granular / trigonal	in pegmatites & granites / beryl, tourmaline, lepidolite & o. / —	ROSE QUARTZ SiO$_2$ / San Miguel on Rio Jequetinonha, Brazil

No. 529−536

No.	Colour Streak	Lustre Transparency Refractive Index	Hardness Specific Gravity	Cleavage Fracture & o. Phys. Props.	Common Form, Aggregates Crystalline Syst.	Occurrence Assoc. Minerals Similar Minerals	Name & Chem. Formula Origin of Specimen
537	yellow-green, differently coloured — white	greasy, dull opaque —	7 / 2.5 — 2.6	— uneven, splintery	most finely granular, massive, concentrically arranged layers, nodules, pebbles, irregular masses (trigonal) amorphous	chalcedony with extraneous admixtures — cf. No. 503	JASPER SiO_2 Kabamby, Madagascar
538	variegated — white	vitreous, greasy transluc., opaque 1.55	7 / 2.5 — 2.6	— conchoidal	most finely granular, massive, compact, stratified, nodules, pebbles, irregular masses (trigonal) amorphous	chalcedony of layered structure, fine-shelled; filling in amygdaloidal volcanic rocks — cf. No. 504	AGATE SiO_2 Levin, Czechoslovakia
539	blood-red, yellow-red — white, reddish	greasy transluc. 1.55	7 / 2.5 — 2.6	— conchoidal	massive, compact, reniform, botryoidal, nodules, pebbles, irregular masses trigonal	cf. No. 422	Chalcedony variety CORNELIAN SiO_2 Brazil
540	apple-green — white	vitreous, greasy transluc. 1.55	7 / 2.6	— conchoidal	massive, reniform, shelled, disseminated, crypto-crystalline trigonal	in nickel ore deps. — prehnite, cf. No. 297	Chalcedony variety CHRYSOPRASE SiO_2 Szklary, Poland
541	yellow, red, brown — white	adamantine transpar., transluc. 2.6 to 2.9	6.5 / 4.2 — 4.3	good brittle	finest, needle-shaped, capillary crystals; as inclusions in rock crystal & smoky quartz tetragonal	as inclusions in quartz, in alpine cracks — cf. No. 347	SAGENITE TiO_2 (Rutile) Modriach, Styria, Austria
542	red-brown, metallic iridescence — white	vitreous transluc. 1.54	5.5 — 6 / 2.6	— uneven	compact, finely granular, massive, slaty trigonal	quartz with embedded lamellae of mica or haematite — cf. No. 289	Quartz variety AVENTURINE SiO_2 (Aventurine Quartz) Mariazell, Styria, Austria
543	gold-brown, blue-green, yellowish — white	silky, dull transluc., opaque —	5.5 — 6 / 3.2 — 3.3	very good fibrous	agg. of parallel fine fibres, zonally curved, platy monoclinic	fibrous riebeckite asbestos, partly silicified & decomposed — cf. No. 488	CROCIDOLITE $Na_2Fe_3Fe_2[(OH,F)Si_4O_{11}]_2$ (Riebeckite Asbestos, Tiger's Eye, Cat's Eye) South Africa
544	grey, brown, blue, violet, shining, grey with a rich play of colour — white	pearly, vitreous transluc. 1.54	6 — 6.5 / 2.7	very good uneven, brittle	thick-tabular, compact, spathic, granular, massive triclinic	constituent in basic igneous rocks & in crystalline schists — —	LABRADORITE isomorphous mixture of $Na[AlSi_3O_8]$ & $Ca[Al_2Si_2O_8]$ (Felspar Family) Kangek, Greenland

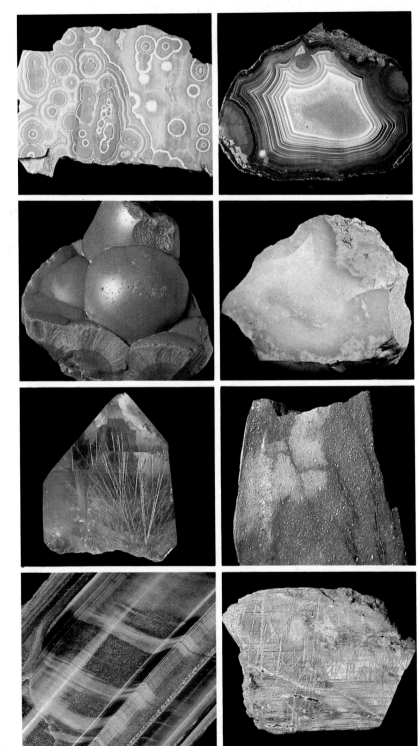

No.	Colour / Streak	Lustre / Transparency / Refractive Index	Hardness & o. / Specific Gravity	Cleavage / Fracture & o. / Phys. Props.	Common Form, Aggregates / Crystalline Syst.	Occurrence / Assoc. Minerals / Similar Minerals	Name & Chem. Formula / Origin of Specimen
545	green, bluish-green / white	vitreous, pearly / transluc. / 1.53	$\frac{6}{2.5}$	very good / uneven, brittle	short-columnar, tabular crystals; compact, granular, spathic / triclinic	constituent in igneous rocks, in pegmatites / — / —	Microcline variety AMAZONITE $K[AlSi_3O_8]$ (Amazonstone) / Pikes Peak, Colorado, USA
546	sky-blue, greenish-blue / white	greasy / opaque / 1.61	$\frac{5-6}{2.6-2.9}$	— / conchoidal, brittle	cryptocrystalline, compact, reniform, massive, conchoidal, botryoidal, crusts, veins, coatings / triclinic	as filling of cracks in siliceous schists & sandstone / chalcedony / lazurite	TURQUOISE $CuAl_6$ $[(OH)_2PO_4]_4 \cdot$ $4H_2O$ (Callaite) / Anatoly, Turkey
547	sky-blue, greenish-blue / light blue	greasy, dull / opaque / 1.56	$\frac{5.5}{2.3-2.4}$	— / conchoidal, brittle	mostly compact, massive, finely granular / cubic	in metamorphic limestones, with disseminated pyrite, in volcanic ejecta / — / sodalite, nosean, hauyne, lazulite	LAZURITE $(Na,Ca)_8$ $[(SO_4,S,Cl)_2$ $(AlSiO_4)_6]$ (Lapis Lazuli) / Baikal on Lake Baikal, USSR
548	green, striated / light green	vitreous, silky / opaque / 1.7 to 1.9	$\frac{3.5-4}{4}$	— / jointing, fibrous, brittle	compact, reniform, botryoidal, mammilliform, agate-like, striated / monoclinic	in weathering zone of copper deps. / — / chrysocolla, pseudomalachite, cf. No. 273	MALACHITE Cu_2 $[(OH)_2CO_3]$ / Katanga, Zaïre
549	green, grey-green / white	greasy, dull / transluc. / 1.62	$\frac{5.5}{2.9-3}$	— / very massive & tenacious (tough)	cryptocrystalline, slaty, compact, massive / monoclinic	massive actinolite or anthophyllite; in loose blocks & pebbles of serpentinized gabbro / — / jadeite, cf. No. 287	NEPHRITE $(Mg,Fe)_7$ $[OHSi_4O_{11}]_2Na_2Ca$ $(Mg,Fe)_{10}$ $(OH)_2O_2Si_{16}O_{44}]$ (Amphibole Family) / New Zealand
550	greenish-white, green, grey, yellow-ish / white	vitreous, greasy / transluc. / 1.65	$\frac{6.5-7}{3.2-3.3}$	— / uneven, very tenacious (tough)	compact, massive, granular, fibrous, nodules, pebbles, cryptocrystalline / monoclinic	inserts in crystalline schists / — / nephrite, cf. No. 301	JADEITE $NaAl[Si_2O_6]$ (Pyroxene Family) / China
551	dark green / white	vitreous, dull / transpar. / 1.48 to 1.5	$\frac{5.5}{2.3-2.4}$	— / conchoidal, sharp-edged	irregular pieces, nodular, grooved chippings with sculptured surface / amorphous	tektite glass with schlieric (streaked) texture, in sands / — / —	MOLDAVITE Besednice, Czechoslovakia
552	honey-yellow, orange, brown-ish, whitish / white	greasy / transpar., transluc. / 1.54	$\frac{2-2.5}{1-1.1}$	— / conchoidal, brittle, combustible	pebbles, rounded grains, plates, stratified, drop-shaped, dissemi-nated / amorphous	fossil resin, in sedimentary rocks / — / cf. No. 140	AMBER $C_{12}H_{20}O$ (Succinite) / Warnemünde, Mecklenburg, Germany

No. 545—552   Table 6

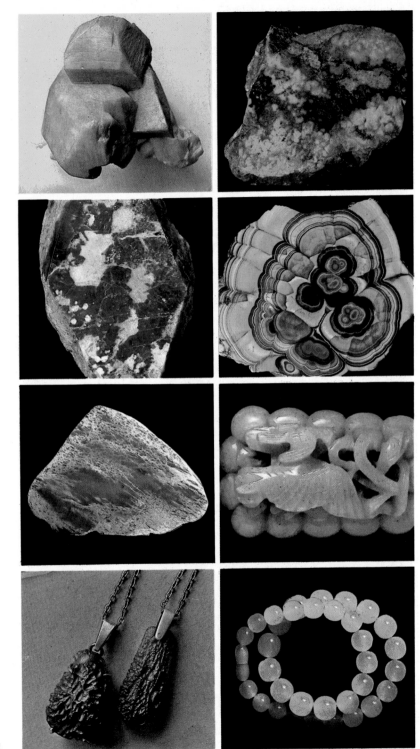

ROCKS a) IGNEOUS ROCKS

No.	Mode of Formation	Structure	Main Constituents	Accessory Minerals	Colours	Specific Gravity / Remarks	Name / Origin of Specimen
553	magmatic plutonic rock (acidic)	coarse- to fine-grained	quartz, orthoclase, microcline, plagioclase, muscovite, biotite, amphibole, augite	apatite, zircon, topaz, tourmaline, beryl, titanite, rutile, magnetite, ilmenite, haematite, pyrite, monazite, fluorite, cordierite, garnet & o.	white-grey, dark grey, white-reddish, greenish, yellowish, bluish, red, black-white	2.6–2.7 slab structure, very hard, solid	**GRANITE** Giant Mountains, Czechoslovakia
554	magmatic plutonic rock	fine- to coarse-grained, sometimes porphyritic	orthoclase, plagioclase, biotite, amphibole, augite; contains none or very little quartz	apatite, zircon, titanite, manganite, olivine, garnet	grey, dark grey, reddish	2.7–2.9 very hard, solid	**SYENITE** Plauenscher Grund, Germany
555	magmatic plutonic rock	medium- to fine-grained	amphibole, pyroxene, biotite, some quartz, diallage	apatite, titanite, rutile, magnetite, ilmenite, pyrrhotite, pyrite	black-white, grey-green, regularly speckled	2.8–3 very hard, tenacious	**DIORITE** Göda near Bautzen, Germany
556	magmatic plutonic rock (basic)	coarse- to fine-grained	plagioclase, pyroxene, amphibole, biotite, some olivine, enstatite, hypersthene	quartz, apatite, ilmenite, spinel, corundum, magnetite, titanite, pyrrhotite, chromite, pyrite, rutile, garnet	grey-black, white-grey, brownish, greenish, grey-green, reddish, speckled	2.8–3.1 solid, tenacious	**GABBRO** Hof, Bavaria, Germany
557	magmatic extruded rock (basic)	fine-grained, massive, vitreous, with insets, porphyritic, also porous	orthoclase, plagioclase, biotite, amphibole, pyroxene, some quartz	apatite, zircon, titanite, magnetite	grey, yellowish, brown, brown-green, reddish, light speckles	2.5–2.8 not too solid	**PORPHYRY** Tatobity, Czechoslovakia
558	magmatic extruded rock (basic)	fine-grained, very massive, porphyritic	plagioclase, nepheline, leucite, augite, amphibole, melilite, olivine, glass	magnetite, ilmenite, biotite, apatite, hauyne, perovskite, zeolite, aragonite, calcite	black, greenish, dark green, grey	2.8–3.3 very hard, solid, tenacious, columnar jointing	**BASALT** Spechtshausen near Freital, Germany
559	magmatic extruded rock (basic)	fine- to coarse-grained, massive, compact, porphyritic, porous, amygdaloidal	plagioclase, augite, hypersthene, olivine, some glass, ore minerals	chlorite, magnetite, ilmenite, agate, quartz, calcite, zeolite	black, dark grey, brown, reddish, red-brown, grey-green	2.8–3.3 fresh hard & tenacious	Basalt variety **MELA-PHYRE** (Amygdaloid, Mandelstone) Kozákov, Czechoslovakia
560	magmatic extruded rock (acidic)	vitreous, fine-crystalline (lava solidified due to quick cooling)	glass	(only in microscopic quantities) quartz, biotite, oligoclase	black, dark grey, brown	2.3–2.6 brittle, conchoidal fracture, used also as gemstone	**OBSIDIAN** Lipari, Italy

No. 553–560   Table 7

ROCKS b) SEDIMENTARY ROCKS

No.	Mode of Formation	Structure	Main Constituents	Accessory Minerals	Colours	Specific Gravity / Remarks	Name / Origin of Specimen
561	clastic sedimentary rock	coarse-grained, angular pieces cemented by some binding agent	varying, e.g. quartz, calcite, dolomite, ore minerals & o.	varying	different	strongly varying, grains larger than 2 mm	**BRECCIA** (Dolomite & Calcite) Tinos, Greece
562	clastic sedimentary rock	unequal grains, pebbles cemented by binding agent	quartz, quartzite, siliceous schist	clay minerals, limonite, haematite, calcite & o.	varying due to colour of mineral constituents & binding agent	varies strongly, grains larger than 2 mm	**CONGLOMERATE** Bělá, Czechoslovakia
563	clastic sedimentary rock	clastic, medium- to fine-grained, porous, stratified	quartz, some opal	chalcedony, muscovite, felspar, haematite, limonite, zircon, rutile, glauconite & o.	light grey, yellowish, reddish, brown, multi-coloured	$2-2.65$ grain size 0.05 to 2 mm	**SANDSTONE** Vrchlabí, Czechoslovakia
564	clastic sedimentary rock	fine-grained to massive, clastic, stratified, schistose	kaolinite, or o. clay minerals, quartz	muscovite, zircon, rutile, calcite, bituminous substances	grey, blue-grey, white-grey, yellowish, yellow-brown, reddish to black	2.8 grain size less than 0.02 mm	**CLAYSTONE** Kopanina, Czechoslovakia
565	clastic, partly biogenetic rock	massive, porous, layered	calcite, dolomite, quartz, clay minerals	clay minerals, bituminous substances	light to dark grey, yellowish, reddish	$2.6-2.8$ grain size less than 0.02 mm	**MARL** Jáchymov, Czechoslovakia
566	biogenetic, partly clastic sedimentary rock	fine-grained to massive, clastic, layered, schistose	kaolinite & o. clay minerals, quartz	muscovite, calcite, zircon, rutile, bituminous substances	grey, blue-grey, reddish to black	2.8 grain size less than 0.02 mm, easily cleaved	**SLATE CLAY** Handlová, Czechoslovakia
567	chemical partly biogenetic sedimentary rock	porous, layered	calcite, aragonite	clay minerals, quartz, haematite, limonite, bituminous substances	white, yellow, brown, reddish, grey-white, marbled	$2.6-2.8$ —	**TRAVERTINE** Sliač, Czechoslovakia
568	chemical sedimentary rock	oölithic, massive, porous, earthy	alunogel, diaspore, böhmite, hydrargillite, kaolinite	haematite, goethite, lepidocrocite, chlorite, calcite, phosphorite, opal	white, yellow, brown, red-brown, violet, green, grey, speckled	$2.4-2.5$ —	**BAUXITE** Gánt, Hungary

No. 561—568   Table 71

ROCKS c) METAMORPHIC OR ALTERATION ROCKS

No.	Mode of Formation	Structure	Main Constituents	Accessory Minerals	Colours	Specific Gravity / Remarks	Name / Origin of Specimen
569	alteration rock	medium- to small-grained, veinous	quartz, orthoclase, plagioclase, biotite, muscovite, amphibole, pyroxene	apatite, zircon, rutile, garnet, cordierite, silimanite, epidote, pyrite, graphite	grey, grey-white, brown, red-brown, grey-black	2.6 – 2.8 slab jointing	GNEISS Kunštát, Czechoslovakia
570	alteration rock	small- to medium-grained, schistose, veinous	quartz, muscovite, biotite, paragonite	albite, garnet, rutile, staurolite, epidote, tourmaline, kyanite, andalusite, graphite	yellowish, brown, brown-red, grey	2.6 – 3.2 slab jointing	MICA SCHIST Vrchlabí, Czechoslovakia
571	alteration rock	fine-grained, massive, schistose, veinous	quartz, orthoclase, plagioclase, garnet, diopside, hypersthene	apatite, zircon, rutile, kyanite, hercynite, biotite	white, grey, grey-white, dark grey	2.6 – 2.9 —	GRANULITE Gersdorf near Karl-Marx-Stadt, Germany
572	alteration rock	fine-grained, schistose, veinous, massive	quartz, chlorite, sericite, partly albite	rutile, albite, tourmaline, magnetite	greenish, grey-green, grey, grey-white to black	2.7 – 2.8 often with micaceous to silky lustre	PHYLLITE Ankogel, Salzburg, Carinthia, Austria
573	alteration rock	small- to coarse-grained, massive, veinous	calcite, dolomite	quartz, mica, talc, epidote, serpentine, tremolite, forsterite & o.	white, yellowish, bluish, green, grey, red, black, multicoloured	2.6 – 2.8 —	MARBLE Schwarzenberg, Ore Mountains, Germany
574	alteration rock	fine-grained to massive, schistose, veinous, compact	chlorite, sericite, amphibole, actinolite, epidote, albite	quartz, calcite	green, grey-green	2.7 – 2.8 —	GREEN-SCHIST (Greenstone) Kraslice, Czechoslovakia
575	alteration rock (basic)	small- to coarse-grained, schistose, veinous, massive	amphibole, plagioclase	albite, quartz, garnet, apatite, titanite, rutile, epidote, zoisite, biotite, chlorite, diopside	black, grey-black, greenish-black	2.7 – 2.8 tenacious	AMPHI-BOLITE Stupava, Czechoslovakia
576	alteration rock (basic)	schistose, veinous, massive, compact	olivine, serpentine	garnet, bronzite, chromite, amphibole, magnetite, talc, chrysotile (asbestos)	grey-green, green, black-green, grey to black, speckled red	2.6 – 2.7 —	SERPEN-TINE Dobšiná, Czechoslovakia

No. 569 – 576   Table

IDENTIFICATION TABLE ACCORDING TO COLOUR AND HARDNESS

Hardness	Colour					
	colourless	white-whitish	yellow-yellowish	orange	red-reddish-pink	violet
1–2	17, 25, 29–31, 33, 35–39	(1), 2, 17–31, 33, 35, 36–38, 255, 256, 393, 396, 397–399	17, 18–25, 27, 28, 31, 33, 35, 36, 38, 39, 129, 131–133, 252, 313, 396–399	33, 129, 133, 181	19, 21, 22, 28, 31, 33, 36, 39, 181–183, 210	33, 209, 210
2–3	40, 42–46, 48, 50, 139, 309, 312, 402	3–7, 40–50, 139, 140, 211, 309, 310, 312, 364, 380, 402	7, 41–48, 119, 121, 122, 133–136, 138–141, 171, 172, 259, 260, 309–312, 326, 403	133, 135, 140, 141, 169–172, 188, 326	40, 44–49, 135, 171, 172, 177, 184–189, 210, 211, 326, 371, 429	46, 210, 211, 371
3–4	51, 55, 57–59, 61, 66, 67, 69, 71, 142, 151, 232, 321, 401, 405	8, 51, 55–61, 63–72, 146, 151–153, 174, 193, 211, 232, 310, 318, 320, 321, 381, 395, 401, 405, 487	51, 55–57, 59–61, 64, 67–70, 72, 123, 124, 128, 142, 143, 146–153, 174, 175, 193, 232, 270, 310, 317, 318, 320–322, 327, 328, 384, 401, 405, 481, 487, 489	143, 146, 147, 148, 174, 175, 327	51, 57–59, 61, 64, 67, 70–72, 142, 151, 192–196, 211, 232, 270, 320, 324, 406, 489	58, 61, 211, 212
4–5	73, 75–80, 158, 213, 214, 236, 276	9, 12, 73–80, 157, 166, 164, 280, 310, 387, 388, 411	73–80, 156–160, 164, 197, 214, 236, 280, 310, 322, 329, 331, 411	73, 156, 280, 332	73, 75–78, 80, 157, 158, 179, 197, 213, 214, 280, 411	73, 158, 213, 214
5–6	81, 84–89, 91–95, 98, 99, 101–104, 112, 158, 161, 204, 283, 286, 291, 421	10–16, 81, 104, 112, 154, 158, 161, 162, 204, 238, 239, 288, 291, 389–392, 414, 416, 421, 486, 495	81, 84–86, 88–91, 94, 97, 98, 100, 103, 112, 154, 155, 158, 161–163, 198, 235, 238, 283, 286, 291, 336, 337, 341, 415, 417, 459, 486, 488, 495	155, 337	81, 85, 88–90, 100, 102, 103, 112, 158, 162, 180, 198–202, 204, 235, 238, 283, 289, 291, 337, 386, 415, 417, 435, 486, 495	158, 215
6–7	105, 107–109, 112, 161, 205, 236, 355, 419, 531	105–109, 112, 161, 164, 165, 205, 206, 236, 293, 294, 301, 349, 419, 491, 498, 500, 502, 506	106–109, 112, 125, 127, 161, 164–166, 203, 205, 236, 292, 295, 301, 349, 351, 355, 418, 420, 422, 438, 500, 529, 532, 541, 544	355	107, 109, 112, 200, 201, 203, 205–207, 292, 293, 295, 299, 347, 349–351, 355, 418, 420, 500, 502, 526, 536, 539, 541, 544	107, 109, 216, 217, 357, 526, 530, 533, 544
7–8	113–117, 427, 521, 522	113, 115, 116	113–117, 167, 168, 176, 244, 246, 305, 427, 518, 521, 522	117, 176, 521, 522	116, 117, 176, 208, 219, 220, 246, 305, 360, 519–522, 527	208, 219, 220, 244
8–9	428, 512, 515	514	304, 428, 512, 515	515	428, 512, 515, 524	515, 524
10	118, 513		118, 513		118, 513	

Colour				
blue-bluish	green-greenish	brown-brownish	grey	black-blackish
21, 33, 209, 221, 223 – 225, 250	19 – 21, 23, 24, 28, 36, 209, 210, 247 – 256, 313, 398	19, 20, 25, 31, 36, 39, 129, 132, 255, 256, 313, 393, 394, 396, 399, 442	2, 20, 22 – 24, 31, 35, 36, 39, 210, 255, 256, 313, 361, 362, 393, 394, 396 – 399, 441	39, 209, 221, 247, 248, 313, 441, 442
222, 226 – 229, 263, 265	136 – 139, 171, 172, 210, 211, 228, 229, 259 – 267, 311, 400, 402, 403, 444	40, 47, 49, 119, 135, 139 – 141, 169 – 171, 177, 189, 265, 267, 307, 309 – 312, 326, 400, 403, 404, 429, 444	3, 5, 6, 41 – 44, 47, 49, 171, 172, 184 – 186, 210, 211, 222, 228, 307, 309 – 312, 364 – 367, 369 – 380, 400, 402 – 404, 432	5, 6, 169, 177, 184, 186, 307, 310, 365, 367, 372, 373, 379, 380, 400, 402, 404, 429, 430, 432, 444
57, 58, 61, 68, 69, 230, 232 – 234, 243, 265	61, 64, 68, 142, 146, 149, 150, 153, 174, 175, 211, 264, 265, 268 – 273, 275, 314, 317, 324, 327, 407, 452, 489	51, 57, 59 – 61, 68, 69, 71, 72, 123, 128, 142, 143, 146 – 148, 150 – 152, 174, 175, 192, 195, 265, 269, 270, 308, 310, 314 – 322, 324, 325, 327, 328, 401, 406, 447, 449, 481, 487, 489	8, 56 – 61, 64, 66, 67, 69 – 71, 142, 146, 153, 174, 192, 195, 211, 270, 310, 317, 321, 381 – 384, 395, 401, 405 – 407, 431, 446, 451, 452, 481, 487, 489	142, 269, 275, 310, 314, 317, 321, 324, 383, 401, 405 – 407, 431, 445 – 449, 451, 452, 481, 489
73, 79, 236, 276, 280, 282, 338, 411	73, 78, 79, 157 – 160, 213, 214, 269, 276 – 282, 284, 329, 333, 338, 411	79, 156, 157, 197, 269, 310, 329 – 333, 338, 454	9, 12, 74, 76, 79, 80, 160, 164, 179, 236, 280, 310, 330, 331, 333, 385, 387, 388, 411, 443, 453	269, 310, 329, 332, 443, 450, 453, 454, 456, 460
87, 93, 112, 215, 235, 237 – 239, 283, 386, 416, 421, 470, 488, 495, 546, 547	87, 90, 93, 94, 98, 101, 103, 104, 112, 158, 162, 163, 238, 239, 283, 285 – 291, 343, 344, 412 – 415, 467, 488, 495, 545, 546, 551	81, 90, 91, 112, 154, 161 – 163, 180, 198, 200, 201, 283, 288, 291, 334 – 337, 339, 341 – 345, 348, 408, 409, 412, 413, 415, 416, 435, 455, 457, 459, 462, 464, 486, 488, 542	12 – 16, 82 – 87, 89, 92 – 94, 102, 112, 162, 180, 201, 204, 238, 283, 287, 288, 290, 291, 342, 343, 345, 386, 389 – 392, 408, 409, 412 – 417, 421, 435, 443, 486, 495	112, 162, 198, 201, 235, 286, 334, 341 – 344, 409, 413, 416, 433 – 435, 443, 455, 457 – 470
105, 107, 112, 216, 236, 241, 242, 351, 355, 422, 544	108, 112, 165, 166, 216, 292 – 302, 349, 351, 353 – 357, 418, 529, 540	106, 109, 112, 161, 166, 200, 201, 203, 205 – 207, 293, 295, 299, 340, 347, 349 – 357, 420, 426, 438, 473, 477, 491, 500, 503, 534, 541, 544	108, 109, 112, 164, 165, 201, 205, 206, 236, 241, 292, 294, 295, 301, 340, 357, 418 – 420, 422, 425, 426, 438, 473, 491, 498, 503, 504, 544	112, 201, 203, 295, 299, 340, 354, 356, 418, 425, 438, 440, 471 – 474, 476 – 478, 491, 498, 506
114, 117, 244 – 246, 305, 517, 520 – 522	114, 117, 168, 303, 305, 306, 475, 516 – 518, 520 – 522, 528	116, 176, 219, 246, 358 – 360, 427, 520 – 522	113 – 115, 244, 427	219, 246, 358 – 360, 475, 479
428, 512, 514, 515	304, 524	428, 512	428, 512	
118, 513	118, 513	118, 513	118, 513	118

IDENTIFICATION TABLE ACCORDING TO COLOUR AND LUSTRE

Lustre	Colour					
	colourless	white-whitish	yellow-yellowish	orange	red-reddish-pink	violet
metallic		1—5, 7—16, 179, 364, 380, 381, 387, 388, 392	119, 123—125, 127, 128, 198, 235, 384		4, 177, 179, 180, 198, 201, 235, 371, 386, 429, 435	209, 371
adamantine	39, 142, 309, 355, 402	49, 146, 157, 162, 309, 321, 393, 402	39, 129, 131, 142, 143, 146, 147, 156, 157, 162, 171, 172, 175, 188, 197, 198, 203, 235, 270, 309, 321, 326, 327, 337, 355, 393, 438, 459	143, 146, 147, 156, 171, 172, 175, 181, 188, 326, 327, 337, 355	39, 49, 142, 157, 162, 171, 172, 181, 182, 184, 185, 186, 188, 195—198, 203, 210, 235, 270, 324, 326, 337, 347, 355, 406	210
vitreous	25, 29—31, 33, 35—37, 40, 42—46, 48, 50, 51, 55, 57—59, 61, 66, 67, 71, 73, 75—81, 84—89, 91—95, 98, 99, 101—105, 107—109, 113, 115—117, 158, 161, 205, 213, 232, 236, 237, 283, 286, 291, 312, 355, 405, 419, 427	25, 27, 29—31, 33, 35—37, 40, 42—48, 51, 55, 56—61, 64, 66—68, 70—72, 74—81, 83—95, 99—108, 110, 111, 113, 115, 116, 152, 153, 158, 161, 162, 164, 165, 193, 205, 206, 213, 232, 236—239, 280, 291, 293, 294, 301, 312, 349, 398, 405, 414, 416, 419, 498, 500, 501	25, 27, 31, 33, 35, 36, 42—48, 51, 55—57, 59—61, 64, 67, 68, 70, 72—81, 84—86, 88—91, 94, 98, 100, 103, 106, 107, 111, 113, 115—117, 131, 137, 138, 150, 152, 153, 158, 161, 162, 164—168, 176, 193, 205, 232, 236, 238, 246, 252, 280, 283, 286, 291, 292, 295, 301, 304, 305, 312, 317, 322, 328, 331, 332, 349, 351, 355, 398, 405, 417, 418, 420, 422, 427, 428, 529	33, 176, 280, 332, 355	31, 33, 36, 40, 44—48, 51, 57—59, 61, 64, 67, 70—73, 75—78, 80, 81, 85, 88—90, 100, 102, 103, 107, 111, 116, 117, 158, 162, 176, 183, 189, 192—194, 199, 200, 202, 205, 206—208, 213, 219, 232, 238, 246, 280, 283, 289, 291—293, 295, 305, 349—351, 415, 417, 355, 406, 418, 420, 428, 519, 527, 530	33, 46, 58, 61, 73, 158, 208, 209, 213, 214, 215, 217, 219, 244, 357, 544
greasy	17, 35, 36, 39, 40, 43, 44, 57, 61, 69, 81, 85, 86, 93, 109, 142, 146, 158, 237, 276, 286, 291, 427	17, 20, 22, 35, 36, 40, 43, 44, 57, 61, 62, 64, 69, 74, 81, 82, 85, 86, 93, 110, 112, 140, 157, 158, 159, 162, 165, 237, 291, 301, 310, 349, 416, 487, 495, 498	17, 20, 22, 35, 36, 39, 43, 44, 57, 61, 62, 64, 69, 74, 81, 85, 86, 112, 129, 131, 140—142, 146—148, 155, 157—159, 162, 165—167, 171, 172, 175, 176, 203, 235, 286, 291, 301, 327, 331, 349, 427	140, 141, 146, 147, 169, 171, 172, 175, 176, 181, 327	22, 36, 39, 40, 44, 57, 61, 64, 81, 85, 112, 142, 157, 158, 162, 171, 172, 176, 181, 187, 203, 208, 235, 291, 349, 536	61, 158, 208, 214, 217, 244
pearly	31, 40, 50, 51, 55, 57—59, 69, 71, 79, 87, 99, 101—105, 107, 113, 205, 213, 232, 236, 312	20, 21, 23, 27, 31, 40, 47, 49, 51, 55, 57—60, 69—72, 79, 83, 87, 99—105, 107, 113, 152, 164, 193, 205, 211, 213, 232, 236, 288, 293, 294, 312, 395, 397, 495	20, 21, 23, 27, 31, 47, 51, 55, 57, 59, 60, 69, 70, 72, 79, 100, 103, 107, 113, 131, 132, 136—138, 150, 152, 164, 193, 205, 232, 236, 311, 312, 317, 328, 397, 403, 420, 495		21, 31, 40, 47, 49, 51, 57—59, 70—72, 100, 102, 103, 107, 189, 193, 194, 200, 205, 210, 211, 213, 232, 293, 395, 415, 420, 495	58, 209, 210, 211, 213, 544
silky	31, 45, 48, 79, 80, 84, 88, 139, 312	27, 28, 31, 41, 45, 48, 68, 72, 79, 80, 83, 84, 88, 134, 139, 206, 255, 312, 398, 414	27, 28, 31, 41, 45, 48, 68, 72, 79, 80, 84, 88, 132—134, 139, 148, 163, 255, 312, 336, 398, 415	133	28, 31, 45, 48, 72, 80, 88, 206, 415	216
dull	25, 30, 37, 38, 42, 43, 48, 69, 92, 113, 405	18—21, 23—26, 30, 37, 38, 41—43, 48, 56, 63, 65, 69, 70, 82, 92, 96, 97, 113, 134, 154, 160, 249, 310, 380, 396, 399, 405, 486, 501	18—21, 23—25, 38, 41—43, 48, 56, 69, 70, 97, 113, 133—145, 149, 150, 154, 156, 160, 305, 317, 329, 335, 396, 399, 405, 422, 428	133, 135, 143, 156, 170	19, 21, 48, 70, 135, 154, 184, 186, 187, 194, 201, 305, 428, 435, 486, 489, 499	209

Colour				
lue-bluish	green-greenish	brown-brownish	grey-greish	black-blackish
), 221, 222, 5	209	119, 123, 128, 180, 198, 201, 307, 429, 435, 449, 454, 455	5, 8, 9, 12 − 16, 179, 180, 201, 222, 307, 361, 362, 364, 365 − 367, 369 − 388, 392, 425, 431, 432, 435, 441, 446, 451	5, 198, 201, 209, 221, 235, 307, 365, 367, 372, 373, 376, 379, 380, 383, 425, 429 − 435, 440, 441, 446, 449 − 451, 453 − 455, 458, 460, 476
5, 235, 355	142, 146, 157, 162, 171, 172, 175, 266, 269, 270, 324, 327, 355, 402, 406	39, 49, 129, 142, 143, 146, 147, 157, 162, 171, 175, 198, 203, 269, 270, 309, 316, 321, 324, 326, 327, 337, 340, 347, 355, 393, 438, 459, 462	39, 49, 142, 146, 171, 172, 184 − 186, 195, 210, 270, 309, 321, 340, 393, 402, 406, 438	39, 142, 162, 184, 186, 198, 203, 235, 269, 321, 324, 340, 402, 406, 438, 459, 462, 469
, 57, 58, 61, , 73, 79, 87, , 105, 107, 7, 209, 226 − 0, 232, 236 − 9, 241, 242, 4 − 246, 250, 3, 280, 305, 1, 355, 416, 2, 428, 544	36, 61, 64, 68, 73, 78, 79, 90, 93, 94, 98, 101, 103, 104, 108, 117, 138, 153, 158, 162, 165, 166, 168, 206, 209, 213, 228, 236 − 239, 248, 250, 252, 253, 254, 257, 258, 262, 263, 266 − 269, 271 − 273, 275, 277, 280, 283 − 286, 289 − 306, 313, 314, 317, 333, 343, 344, 349, 351, 353, 355, 356, 357, 398, 406, 407, 412 − 415, 418, 452, 529, 545, 551	25, 31, 36, 40, 47, 51, 57, 59 − 61, 68, 71, 72, 79, 81, 90, 91, 106, 111, 116, 150, 152, 161, 162, 166, 176, 189, 192, 200, 205 − 207, 219, 246, 267, 269, 283, 291, 293, 295, 299, 312 − 314, 317, 322, 328, 331 − 333, 340, 342, 343 − 345, 349 − 353, 355 − 359, 394, 404, 408, 409, 412, 413, 415, 416, 420, 426 − 428, 444, 447, 464, 473, 477, 534, 544	31, 35, 36, 42 − 44, 56 − 61, 64, 66, 67, 70, 71, 76, 79, 80, 83 − 87, 89, 92 − 94, 102, 107, 108, 111, 113, 153, 164, 165, 192, 205, 206, 228, 236 − 238, 241, 244, 280, 283, 292, 294, 295, 301, 312, 313, 317, 331, 340, 343, 345, 357, 394, 398, 404 − 409, 412 − 420, 422, 426 − 428, 452, 473, 491, 498, 501, 534, 544	162, 209, 219, 246, 269, 273, 275, 286, 295, 299, 313, 314, 317, 332, 340, 342 − 344, 356, 358, 359, 404 − 407, 409, 413, 416, 418, 444, 447, 448, 452, 456, 457, 461, 464, 465, 467, 468, 470, 471, 473 − 475, 477 − 479, 491, 498
, 61, 69, 93, 5, 237, 242, 4, 245, 265, 6, 546, 547	20, 22, 36, 61, 64, 93, 112, 142, 146, 157, 158, 159, 162, 165, 166, 171, 172, 175, 237, 248, 262, 265, 267, 268, 275, 276, 279, 282, 286, 291, 297, 301, 303, 327, 329, 338, 349, 443, 489, 495	20, 36, 39, 40, 57, 61, 69, 81, 112, 129, 140 − 142, 146 − 148, 157, 162, 166, 169, 171, 175, 176, 203, 265, 267, 291, 310, 327, 331, 334, 341, 342, 345, 349, 359, 416, 424, 427, 454, 455, 464, 473, 487	20, 35, 36, 39, 43, 44, 57, 61, 64, 82, 85, 86, 93, 112, 142, 146, 165, 171, 172, 237, 244, 301, 331, 416, 424, 427, 443, 473, 487, 495, 498	39, 112, 142, 162, 169, 203, 235, 275, 286, 310, 329, 341, 342, 359, 416, 424, 443, 454, 455, 457, 460, 464, 468, 469, 472 − 474, 478, 479, 489, 498
, 57, 58, 69, , 87, 105, 7, 209, 225, 2, 236, 243, 0, 544	20, 21, 23, 79, 101, 103, 104, 136, 138, 209, 211, 213, 236, 248, 250, 251, 253, 254, 257, 258, 261, 279, 288, 293, 294, 311, 313, 317, 333, 400, 403, 412, 413, 415, 452, 495, 545	20, 31, 40, 47, 49, 51, 57, 59, 60, 69, 71, 72, 79, 132, 150, 152, 189, 200, 205, 288, 293, 311, 312, 313, 317, 328, 333, 403, 408, 412, 413, 415, 420, 444, 447, 544	20, 23, 31, 40, 57 − 60, 70, 71, 79, 83, 87, 102, 107, 113, 164, 205, 210, 211, 236, 294, 311, 312, 313, 317, 345, 395, 397, 400, 403, 408, 412, 413, 415, 420, 452, 495, 544	209, 313, 317, 400, 413, 444, 447, 452
, 79, 216, 3, 225, 272	28, 68, 79, 139, 163, 206, 216, 255, 273, 290, 398, 413 − 415, 488	31, 68, 72, 79, 132, 139, 148, 163, 206, 255, 312, 336, 408, 413, 415, 462, 488	31, 41, 79, 80, 83, 84, 206, 312, 398, 408, 413 − 415, 451	273, 413, 451, 456, 462
, 69, 209, 0, 265, 305, 2, 428, 501, 7	19 − 21, 23, 24, 160, 209, 247, 249, 251, 254, 264, 265, 273, 278, 282, 287, 305, 317, 400, 406, 407, 443, 475, 488, 489	19, 20, 25, 69, 135, 143, 150, 154, 170, 201, 265, 307, 310, 317, 330, 335, 336, 358, 394, 396, 399, 404, 409, 424, 426	19, 20, 23, 24, 41 − 43, 56, 70, 82, 92, 113, 160, 184, 186, 201, 307, 317, 365, 366, 378, 380, 394, 396, 399, 400, 409, 422	184, 186, 201, 209, 273, 307, 310, 317, 358, 365, 380, 400, 409, 424, 425, 433, 441 − 443, 446, 450, 453, 456, 458, 460

191

IDENTIFICATION TABLE ACCORDING TO COLOUR AND CLEAVAGE

Cleavage	Colour					
	colourless	white-whitish	yellow-yellowish	orange	red-reddish-pink	violet
very good	25, 31, 33, 35, 42, 44—46, 48, 49, 51, 55, 57—59, 66, 71, 73, 79, 80, 88, 91, 94, 99, 101—105, 107, 108, 115, 117, 118, 142, 205, 213, 236, 309, 312, 405	2, 4, 7, 13, 20, 21, 23, 25, 27, 31, 33, 35, 41, 42 ,44—48, 35, 41, 42, 44—48, 51, 55, 57—60, 66, 70—72, 79, 80, 83, 88, 91, 94, 99—105, 107, 108, 115, 153, 164, 205, 206, 211, 213, 236, 238, 280, 293, 309, 312, 364, 380, 381, 395, 397, 405, 414, 416, 500	20, 21, 23, 25, 27, 31, 33, 35, 41, 42, 44—48, 51, 55, 57, 59, 60, 70, 72, 73, 79, 80, 88, 91, 94, 100, 103, 107, 115, 117, 118, 123, 131, 132, 136—138, 142, 149, 150, 153, 164, 168, 197, 205, 235, 236, 238, 252, 260, 280, 292, 295, 309, 311, 312, 317, 328, 331, 337, 397, 403, 405, 415, 420, 488, 500	33, 280, 337	21, 31, 33, 44—49, 51, 57—59, 70—73, 80, 88, 100, 102, 103, 107, 117, 118, 142, 182, 183, 186, 189, 192, 194, 197, 199, 205, 206, 210, 211, 213, 235, 238, 280, 292, 293, 295, 337, 395, 415, 420, 500	33, 46, 58, 73, 107, 209 210, 211 213, 544
good	30, 40, 50, 67, 75, 76, 84, 87, 93, 116, 232, 283, 286, 402	8, 10—12, 15, 28, 30, 40, 56, 67, 68, 74—76, 84, 87, 90, 93, 116, 146, 152, 165, 193, 232, 249, 255, 286, 288, 393, 398, 402	28, 56, 67, 68, 74—76, 84, 90, 116, 133, 143, 146, 152, 156, 163, 165, 175, 188, 193, 203, 232, 255, 283, 335, 393, 398, 417, 418, 459, 481, 529	133, 143, 146, 156, 170, 175, 181, 188	28, 40, 67, 75, 76, 90, 116, 181, 184, 185, 188, 193, 196, 200, 203, 232, 283, 347, 350, 417, 418, 530	215, 24 357, 53
clear	37, 38, 43, 116, 158, 286, 421	10, 15, 16, 37, 38, 43, 64, 116, 134, 157, 158, 162, 165, 286, 294, 349, 392, 421	38, 43, 64, 116, 128, 134, 157, 158, 162, 163, 165, 172, 304, 322, 349	172	64, 116, 157, 158, 162, 172, 195, 349	158, 2
poor	17, 61, 77, 78, 81, 86, 291, 355, 427	17, 61, 77, 78, 81, 82, 86, 106, 291, 320, 321, 387, 388	17, 61, 77, 78, 81, 86, 106, 124—127, 166, 171, 198, 246, 291, 305, 320, 321, 332, 355, 384, 427, 438	171, 332, 355	61, 77, 78, 81, 171, 198, 246, 291, 305, 355, 386	61
none	29, 36, 39, 69, 85, 89, 92, 95, 98, 109, 113, 114, 139, 161, 239, 276, 419	1, 3, 5, 9, 14, 18, 19, 22, 24, 26, 29, 36, 62, 63, 65, 69, 85, 89, 92, 95—97, 110—113, 139, 140, 154, 159—161, 179, 301, 310, 318, 390, 391, 396, 399, 419, 486, 487, 495, 498, 501	18, 19, 22, 24, 36, 39, 62, 69, 85, 89, 97, 98, 111—114, 119—122, 124, 125, 127, 129, 135, 139—141, 147, 148, 154, 155, 159—161, 167, 176, 270, 301, 326, 330, 336, 341, 351, 396, 399, 422, 428, 438, 486, 487, 489, 491, 495, 499, 501, 518	129, 135, 140, 141, 147, 169, 176, 326	19, 22, 36, 39, 85, 89, 111, 112, 135, 154, 176, 177, 179, 180, 187, 201, 202, 207, 208, 219, 270, 289, 324, 326, 351, 371, 406, 428, 429, 486, 489, 495, 499, 519, 527, 536	208, 21 219, 371

Colour				
blue-bluish	**green-greenish**	**brown-brownish**	**grey**	**black-blackish**
21, 33, 57, 58, 73, 79, 105, 107, 117, 118, 209, 221, 224 – 226, 229, 233 – 236, 238, 243, 250, 280, 416, 470	20, 21, 23, 73, 79, 94, 101, 103, 108, 117, 118, 136, 138, 142, 146, 149, 150, 153, 168, 206, 209, 211, 213, 225, 238, 248, 250, 252 – 254, 257 – 261, 266, 279, 280, 284, 292, 293, 295, 296, 311, 313, 314, 317, 403, 414, 415, 444, 452, 467, 488, 545	25, 31, 47, 49, 51, 57, 59, 60, 71, 72, 79, 91, 123, 132, 142, 146, 150, 189, 192, 205, 206, 279, 293, 295, 309, 311 – 317, 328, 331, 337, 346, 403, 408, 415, 416, 420, 444, 447, 449, 454, 455, 488, 500, 544	13, 20, 23, 31, 35, 41, 42, 44, 49, 57 – 60, 66, 70, 71, 79, 80, 83, 94, 102, 107, 108, 118, 142, 146, 153, 164, 186, 192, 205, 206, 210, 211, 236, 238, 280, 292, 295, 309, 311 – 313, 317, 331, 361, 362, 364 – 366, 367, 372, 380, 381, 395, 397, 403, 405, 408, 411, 414 – 416, 420, 432, 441, 446, 451, 452, 544	118, 142, 186, 209, 221, 235, 295, 313, 314, 317, 367, 372, 380, 405, 416, 432, 434, 441, 444 – 447, 449, 451, 452, 454, 455, 468, 470
?, 87, 93, 215, ?, 87, 93, 215, ?2, 223, 230, ?, 237, 244, ?3	28, 68, 87, 90, 93, 104, 163, 165, 175, 249, 251, 255, 262, 263, 269, 272, 273, 283, 285, 286, 288, 290, 299, 333, 344, 357, 398, 400, 402, 407, 413, 418, 529	40, 68, 90, 116, 143, 152, 156, 163, 170, 175, 288, 299, 327, 333, 335, 344, 345, 347, 350, 354, 357, 393, 394, 404, 413, 459, 462, 481	8, 12, 15, 56, 67, 74, 76, 84, 87, 93, 165, 184, 185, 222, 244, 283, 288, 290, 345, 357, 385, 393, 394, 398, 400, 402, 404, 407, 413, 417, 418, 481	184, 203, 269, 273, 286, 299, 344, 400, 402, 404, 407, 413, 418, 448, 453, 459, 462, 465, 467, 472, 474, 481
..6, 421	64, 157, 158, 162, 163, 165, 172, 216, 281, 286, 294, 304, 343, 349, 412, 466	116, 128, 157, 162, 163, 308, 322, 340, 342, 343, 349, 358, 412, 426, 466, 467	15, 16, 43, 64, 165, 172, 195, 294, 340, 342, 343, 374 – 378, 389, 392, 412, 421, 425, 426, 466	162, 286, 340, 342, 343, 358, 376, 425, 440, 466, 469
61, 227, 228, 246, 338, 355, 386	61, 78, 166, 171, 267, 271, 277, 291, 305, 306, 338, 355	61, 81, 106, 166, 171, 198, 246, 267, 291, 320, 321, 323, 332, 334, 338, 339, 348, 355, 427, 438, 464	61, 82, 86, 171, 321, 365, 369, 370, 373, 383, 384, 386 – 388, 401, 427, 438	198, 246, 321, 332, 365, 373, 384, 433, 438, 464, 476
9, 114, 239, 241, 242, 245, 265, 351, 422, 428, 495, 501, 517, 546, 547	19, 22, 24, 36, 98, 112, 114, 139, 159, 160, 239, 247, 264, 265, 268, 270, 275, 276, 278, 282, 287, 289, 297, 298, 300 – 303, 324, 329, 351, 353, 356, 406, 443, 475, 489, 495, 517, 518, 551	19, 36, 39, 69, 111, 112, 119, 129, 135, 139 – 141, 147, 148, 154, 161, 169, 176, 180, 201, 207, 219, 265, 270, 307, 310, 318, 319, 324 – 326, 329, 330, 336, 341, 351, 352, 353, 356, 359, 360, 396, 399, 409, 424, 428, 429, 438, 442, 473, 477, 486, 487, 489, 491, 499, 501, 534	5, 9, 14, 19, 22, 24, 36, 59, 85, 89, 92, 111 – 114, 160, 179, 180, 201, 241, 270, 301, 307, 330, 363, 371, 377, 379, 382, 390, 391, 396, 399, 406, 409, 419, 422, 424, 428, 431, 438, 443, 473, 486, 487, 491, 495, 498, 501, 534	5, 39, 112, 169, 201, 219, 275, 307, 310, 324, 329, 341, 356, 359, 379, 406, 409, 424, 429 – 431, 433, 435, 438, 442, 443, 450, 456 – 458, 460, 461, 471, 473, 475, 477 – 479, 483, 489, 491, 498

IDENTIFICATION TABLE ACCORDING TO STREAK AND HARDNESS

Hard-ness	Streak					
	white-whitish	silver-white	yellow-yellowish	orange	pink	red-reddish
1—2	17—39, 132, 224, 248, 252, 255, 256, 393, 394, 396—399		129, 131, 133, 134, 183, 251, 393 129, 131, 133, 134, 183, 251, 393	181, 183	210	182, 210
2—3	39—55, 139—141, 170—172, 189, 211, 227, 229, 262, 309, 310, 311, 312, 319, 323, 398, 402, 403, 444, 484	4, 5, 363	7, 119, 129, 133, 135—138, 169, 188, 259, 260, 267, 326	187, 188	210, 211	177, 184—186, 210
3—4	56—73, 142, 145—147, 151—153, 174, 190—194, 211, 212, 232—234, 236, 270, 272, 310, 315, 317, 318, 320, 321, 325, 329, 330, 395, 401, 405, 445, 447, 452, 481, 485, 487, 489, 490		142, 143, 148—150, 175, 191, 269, 270, 315—317, 320, 322, 324, 325, 327, 329, 445, 481, 485	406	211	195, 196
4—5	74—84, 157—160, 164, 213, 214, 236, 276, 280—282, 310, 331, 338, 411	9	156, 197, 269	156, 197, 332		454
5—6	85—105, 112, 155, 161—164, 199, 200, 202, 204, 215, 235, 237—239, 283, 286, 287—291, 333, 337, 339, 343—345, 348, 408, 412—417, 421, 459, 461, 463, 466, 468, 470, 486, 495, 543, 545, 546, 551		154, 198, 235, 334—336, 348, 409, 462, 482, 488			201, 435, 436, 437
6—7	106—114, 161, 164—166, 200, 205—207, 216, 217, 219, 236, 241, 242, 292—302, 340, 346, 347, 349, 351—357, 418—420, 422, 424, 426, 438, 461, 477, 478, 491, 495, 498, 500, 503, 504, 525, 526, 528—530, 534, 536, 544		438, 499			201, 203, 207
7—8	115—117, 167, 168, 176, 208, 220, 244—246, 303, 305, 306, 355, 358—360, 427, 479, 516—523, 525, 527					
8—9	304, 428, 512, 514, 515, 524					
10	118, 513					

			Streak			
blue-bluish	green-greenish	brown-brownish	grey	green-black	grey-black	black
9, 221, 223, 5	225, 247, 249 – 251, 253, 254, 313, 362	442	2, 39, 361, 362, 441			221, 361, 441, 442
6, 228, 265	136, 228, 249, 257 – 267, 400, 404	267, 307, 326	3, 4, 7, 39, 172, 364, 365, 367, 369, 370, 372, 374, 383, 400, 404, 444		307, 366, 375	222, 367, 371, 373, 374, 376 – 380, 429, 430, 432
0, 243, 265, 1	123, 243, 264, 265, 268, 269, 271, 273, 275, 314, 448, 449	142, 191, 315, 316, 324, 328, 406, 445, 451	8, 270, 325, 381, 407, 447, 452, 481, 489, 490	124	123, 128, 308, 382, 446	124, 379, 384, 431, 446, 449, 483
7, 278	269, 277, 279, 284, 443, 456	279, 332, 453, 454, 460	385, 388, 454		12, 387	179, 379, 388, 443, 450, 453, 456, 460
8, 547	285, 342, 443, 467, 470	154, 180, 201, 334, 335, 341, 348, 386, 409, 434 – 437, 455, 457, 458, 460 – 462, 464, 465, 468, 469, 482, 488	215, 333, 337, 343, 344, 459, 463, 464, 466 – 468, 470, 488	342	10, 11, 13, 14, 16, 387, 390 – 392	15, 180, 342, 379, 386, 389, 433, 443, 455, 458, 460, 469
	474, 478	201, 203, 341, 435, 436 – 438, 440, 461, 472, 473, 476	295, 340, 349, 351 – 353, 356, 473, 474, 491	125, 127	471	379, 425, 440
	475		358, 478			

Trans-parency	Colour					
	colourless	white-whitish	yellow-yellowish	orange	red-reddish-pink	violet
transparent	17, 25, 29 − 31, 33, 35, 36, 39, 40, 42 − 46, 50, 51, 55, 57 − 59, 61, 66, 71, 73, 75 − 79, 85 − 89, 91, 93 − 95, 98, 99, 101 − 104, 107 − 109, 113, 115, 116 − 118, 142, 161, 205, 213, 232, 236, 237, 281, 283, 286, 291, 309, 312, 355, 402, 405, 419, 427	17, 25, 29, 30, 31, 33, 35, 36, 40, 42 − 47, 51, 55 − 59, 61, 66, 71, 72, 75 − 79, 85 − 91, 93 − 95, 99, 101 − 104, 107, 108, 110, 113, 115, 116, 140, 153, 157, 161, 162, 165, 205, 211, 213, 232, 236 − 238, 281, 291, 294, 309, 312, 321, 349, 393, 402, 405, 416, 419	17, 25, 31, 33, 35, 36, 39, 42 − 47, 51, 55, 56, 57, 59, 61, 72, 73, 75 − 79, 85, 86, 88 − 91, 94, 98, 103, 107, 113, 115, 116 − 118, 129, 131, 136, 140, 142, 153, 157, 161, 162, 165, 167, 172, 176, 188, 198, 205, 232, 236, 238, 246, 252, 281, 283, 286, 291, 304, 305, 309, 311, 312, 321, 328, 349, 355, 393, 403, 405, 427, 428, 518, 529	33, 129, 140, 172, 176, 181, 188, 355	31, 33, 36, 39, 40, 44 − 47, 51, 57 − 59, 61, 71 − 73, 75 − 78, 85, 88 − 90, 102, 103, 107, 116 − 118, 142, 157, 162, 172, 176, 181, 183 − 185, 188, 189, 198, 199, 202, 205, 208, 211, 213, 219, 232, 238, 246, 281, 283, 291, 305, 347, 349, 355, 428, 519, 527, 530	33, 46, 58, 61, 73, 208, 211, 213, 217, 219, 244, 281, 357, 530
translucent	17, 25, 29, 30, 31, 33, 35 − 40, 42 − 46, 48, 51, 55, 57 − 59, 61, 66, 67, 71, 73, 75 − 81, 84 − 89, 91 − 94, 99, 101 − 105, 107, 108, 113, 115 − 118, 139, 142, 161, 205, 213, 232, 236, 237, 276, 281, 283, 286, 291, 309, 312, 355, 402, 405, 427	17, 18, 20, 23, 25, 27 − 31, 33, 35 − 38, 40 − 49, 51, 55 − 61, 63, 66 − 68, 70 − 72, 74 − 81, 83 − 94, 99 − 108, 110 − 113, 115, 116, 134, 139, 140, 146, 152, 153, 157, 159, 161, 162, 165, 193, 205, 206, 211, 213, 232, 236 − 239, 249, 255, 256, 272, 280, 281, 288, 291, 294, 301, 309, 312, 321, 349, 393, 395, 397 − 399, 402, 405, 414, 416, 486, 487, 495, 498, 501	17, 18, 23, 25, 27, 28, 31, 33, 35, 36, 38, 39, 41 − 48, 51, 55 − 57, 59 − 61, 67, 68, 70, 72 − 81, 84 − 86, 88 − 91, 94, 100, 103, 106, 107, 111 − 113, 115 − 118, 129, 131 − 134, 136 − 143, 146 − 150, 152, 153, 155 − 157, 159, 161 − 163, 165, 166, 168, 171, 172, 175, 176, 193, 197, 198, 203, 205, 232, 235, 236, 246, 248, 252, 255, 256, 270, 272, 280, 281, 283, 286, 291, 292, 295, 301, 304, 305, 309, 311, 312, 317, 321, 326, 328, 331, 332, 337, 349, 351, 355, 393, 397 − 399, 403, 405, 415, 417, 418, 420, 422, 427, 428, 438, 486 − 489, 495, 501	33, 129, 133, 140, 141, 143, 146, 147, 156, 170 − 172, 175, 176, 181, 280, 326, 332, 337, 355	28, 31, 33, 36, 39, 40, 44 − 49, 51, 57 − 59, 61, 67, 70 − 73, 75 − 78, 80, 81, 85, 88 − 90, 100, 102, 103, 107, 111, 112, 116 − 118, 142, 157, 162, 171, 172, 176, 181 − 186, 189, 192 − 194, 196 − 200, 202, 203, 205, 206, 208, 210, 211, 213, 219, 232, 235, 238, 246, 270, 280, 281, 283, 289, 291, 292, 295, 305, 324, 326, 337, 347, 349, 350, 351, 355, 395, 415, 417, 418, 420, 428, 486, 489, 495, 519, 536.	33, 46, 58, 61, 73, 208 − 211, 213, 215 − 217, 219, 244, 281, 317, 544
opaque		1 − 5, 7 − 16, 19 − 22, 24, 26, 62, 63, 65, 69, 82, 92, 96, 97, 103, 111, 112, 154, 160, 179, 239, 249, 293, 310, 318, 320, 364, 380, 381, 387, 388, 392, 396, 399, 486, 487, 500	7, 19 − 22, 24, 62, 69, 97, 103, 111, 112, 118, 119, 123 − 125, 127, 128, 135, 141, 142, 149, 154, 155, 160, 166, 197, 203, 235, 292, 320, 322, 326, 329, 330, 332, 335 − 337, 341, 384, 396, 399, 415, 418, 420, 438, 459, 481, 486 − 489, 491, 499, 500	135, 141, 169, 326, 332, 337	4, 19, 21, 22, 103, 111, 112, 118, 135, 142, 154, 177, 179, 180, 184, 187, 189, 195, 197, 201, 203, 207, 219, 235, 292, 293, 324, 326, 337, 350, 371, 386, 406, 415, 418, 420, 429, 435, 486, 489, 499, 500	209, 215, 219, 371

		Colour		
blue-bluish	**green-greenish**	**brown-brownish**	**grey**	**black-blackish**
, 57, 58, 61, », 79, 87, 93, *7, 117, 118, 29, 232, 236— 18, 244, 246, -0, 305, 355, 6, 428, 470, 7 -	36, 61, 73, 78, 79, 90, 93, 94, 98, 101, 103, 104, 108, 117, 118, 136, 142, 153, 157, 162, 165, 172, 211, 213, 236— 238, 250, 252—254, 257—259, 262, 266, 268, 269, 277, 281, 283—286, 290, 291, 294, 299, 300, 302— 306, 311, 313, 344, 349, 355, 357, 402, 403, 407, 444, 517, 518, 529, 551	25, 31, 36, 39, 40, 47, 51, 57, 59, 61, 71, 72, 79, 90, 91, 116, 129, 140, 142, 157, 161, 162, 176, 189, 198, 205, 219, 246, 269, 283, 291, 299, 309, 311, 312, 313, 321, 328, 344, 347, 349, 355, 357, 359, 393, 403, 416, 427, 428, 444, 477, 534	31, 35, 36, 39, 42, 43, 44, 56—59, 61, 66, 71, 76, 79, 85—87, 89, 93, 94, 102, 107, 108, 113, 118, 142, 153, 165, 172, 184, 185, 205, 211, 236—238, 244, 283, 290, 294, 309, 311— 313, 321, 357, 393, 402, 403, 405, 407, 416, 419, 427, 428, 534	39, 118, 142, 162, 184, 198, 219, 246, 269, 286, 299, 313, 321, 344, 359, 402, 405, 407, 416, 444, 470, 477
8, 57, 58, 61, 8, 73, 79, 87, 8, 105, 107, 17, 118, 209, 16, 223, 225— 80, 232, 235— 89, 244—246, 50, 263, 265, 72, 280, 305, 88, 351, 355, 16, 422, 428, 70, 495, 501, 14	23, 28, 36, 61, 68, 73, 78, 79, 90, 93, 94, 101, 103, 104, 108, 112, 117, 118, 136, 138, 139, 142, 146, 153, 157, 159, 162, 163, 165, 166, 168, 171, 172, 175, 206, 209, 211, 213, 216, 228, 236— 239, 248—259, 261— 263, 265—273, 275— 281, 283—292, 294— 306, 311, 313, 317, 324, 333, 338, 344, 349, 351, 353, 355—357, 398, 402, 403, 407, 412, 414, 415, 444, 452, 467, 488, 489, 495, 545	25, 31, 36, 39, 40, 47, 49, 51, 57, 59—61, 68, 71, 72, 79, 81, 90, 91, 106, 111, 112, 116, 129, 132, 139—143, 146— 148, 150, 152, 157, 161—163, 166, 170, 171, 175, 176, 189, 192, 198, 200, 203, 205, 206, 219, 246, 255, 256, 265, 267, 269, 270, 272, 283, 288, 291, 295, 299, 309, 311—313, 316, 317, 321, 324, 326, 328, 331—334, 337, 338, 340, 341, 344, 345, 347, 349—353, 355—359, 393, 394, 399, 403, 408, 412, 415, 416, 420, 424, 426, 427, 428, 438, 444, 447, 454, 462, 464, 477, 486—489, 501, 544	23, 31, 35, 36, 39, 41— 44, 49, 56—61, 66, 67, 70, 71, 76, 79, 80, 83— 87, 89, 92—94, 102, 107, 108, 111—113, 118, 142, 146, 153, 165, 171, 172, 184—186, 192, 205, 206, 210, 211, 228, 236—238, 244, 270, 280, 283, 290, 292, 294, 295, 301, 309, 311— 313, 317, 321, 331, 340, 345, 357, 393—395, 397—399, 402, 403, 405, 407, 408, 412, 414—418, 420, 422, 424, 426—428, 438, 452, 486, 487, 495, 498, 501, 544	39, 112, 118, 142, 162, 184, 186, 198, 203, 209, 219, 235, 246, 269, 273, 275, 286, 295, 299, 313, 317, 321, 324, 332, 340, 341, 344, 356, 358, 359, 402, 405, 407, 416, 418, 424, 434, 438, 444, 447, 448, 452, 454, 462, 465, 467—470, 477, 489, 498
4, 69, 118, 09, 221, 222, 80, 235, 239, 41—243, 245, 55, 546, 547	19—22, 24, 103, 112, 118, 142, 160, 166, 209, 239, 247—249, 264, 265, 273, 278, 279, 282, 292, 293, 298, 314, 324, 329, 333, 343, 353, 356, 400, 406, 412, 413, 415, 443, 467, 475, 488, 489	19, 20, 60, 111, 112, 119, 123, 128, 135, 141, 142, 154, 166, 169, 180, 189, 201, 203, 207, 219, 265, 293, 307, 310, 314, 316, 318—320, 322, 324, 326, 330, 332— 337, 340—343, 350, 352, 353, 356, 358, 396, 399, 404, 409, 412, 413, 415, 420, 424—426, 429, 435, 438, 442, 449, 454, 455, 459, 462, 464, 473, 481, 486—489, 491, 499, 500	5, 8, 9, 12—16, 19, 20, 24, 82, 92, 111, 112, 118, 142, 160, 179, 180, 184, 195, 201, 222, 241, 292, 307, 319, 330, 340, 343, 361, 362, 364— 367, 369—388, 392, 396, 399, 400, 404, 406, 409, 412, 413, 415, 418, 420, 424—426, 431, 432, 435, 438, 441, 443, 446, 451, 473, 481, 486, 487, 491	5, 112, 118, 142, 169, 184, 201, 203, 209, 219, 221, 235, 246, 273, 307, 310, 314, 324, 329, 332, 340—343, 356, 358, 365, 367, 372, 373, 376, 379, 380, 383, 400, 404, 406, 409, 413, 418, 424, 425, 429—435, 438, 440—443, 446, 448, 449—451, 453—462, 467—469, 471—476, 478, 479, 481, 489, 491

Index of Minerals

Roman figures indicate numbers of minerals; figures in italics refer to text pages.

Saponite 22
Sapphire *23, 164, 166*, 514
Satin Spar 132
Scapolite 87
Scapolite Family 421
Scheelite *15, 25*, 157
Schorl 479
Schroeckingerite 137
Schwazite 446
Scolezite 91
Scorodite 407
Seladonite 247
Selenite 132
Semseyite 376
Sepiolite 24
Serpentine *12*, 489, 576
Shell Marble 484
Siderite *12*, 317, 330
Silicate of Zinc 79
Siliceous Sinter 486
Sillimanite *169*, 206
Silver 5, 6, 363
Silver Fahlerz 382
Silver Glance 365
Silver Telluride 369
Skutterudite 16, 391
Slate Clay 566
Smaltite 16, 388, 391
Smithsonite 280, 411
Smoky Quartz *23, 166*, 534
Soap-Stone 22
Soda Nitre 25
Sodalite *168*, 93, 237
Soddite 150
Spathose Iron 317
Spear Pyrites 127
Specularite 435
Spessartite 176
Sphalerite *24, 27, 28*, 142, 191, 315, 445
Sphene 162, 463
Spinel *166*, 246, 360, 522
Spinel Family 475
Spodumene 108, 529, 530
Sprudelstein 320, 485
Staffelite 159
Stannine 384
Stassfurtite 114
Staurolite 358
Steatite 20
Stellerite 194
Stephanite 373

Stibnite 367, 368
Stilbite 71, 72
Stilpnomelan 314
Stolzite 171, 323
Strontianite 64
Strüverite 471
Struvite *15*, 394
Succinite 140, 552
Sulphur *10, 13, 15, 22, 25*, 129, 130
Sulphuric Pyrites 125, 126
Sunstone 500
Syenite *168*, 554
Sylvanite 2
Sylvite *25*, 35

Table Spar 80
Talc *19, 22, 25, 169*, 20
Tantalite 472
Tarnowitzite 81
Tellurium 364
Tetrahedrite 431, 483
Thomsonite 90
Thorite 332
Thortveitite 340
Thuringite 400
Tiger's Eye 488, 543
Tin Pyrites 384
Tin White Cobalt 388
Tincal 43
Tinsel 131
Tinstone 438, 439
Titanite *15*, 162, 339, 463
Titanoferrite 458
Topaz *19, 23, 24*, 117, 168, 523
Torbernite 258, 261
Tourmaline *11, 18, 22, 28*, 245, 302, 359, 479, 480, 493, 494, 520
Tourmaline Family 208, 220, 359
Travertine 567
Tremolite 414
Tridymite *14*, 113
Troostite 417
Tschermigite 29
Turquoise 546
Tyrolite 250

Ullmannite 12
Uran Ochre 133, 135
Uraninite 443
Uranite 259

Uranium Mica 136, 137, 259
Uranocircite 136, 259
Uvarovite 303

Valentinite 49
Vanadinite 326
Variscite 276
Verdelite 302
Vermiculite 313
Vesuvianite *15*, 242, 351, 352, 353
Violane 215
Vivianite 209

Wad 442
Wagnerite 85
Walchovite 141
Wardite 282
Warwickite 447
Wavellite 68, 272
Wheel Ore 383
Whewellite *10*, 50
White Arsenic 38
White Lead Ore 321, 401
White Nickel 14, 390
Willemite *25*, 283
Wiluite 353
Witherite 56
Wolframite *15, 23*, 455
Wollastonite 80
Wood Opal 155
Wulfenite *15*, 172, 173
Wurtzite 316

Xanthosiderite 336

Yellow Lead Ore 172, 173
Yenite 342

Zaratite 268
Zeolite Family 78, 84, 88, 90, 91, 194, 199
Zinc Blende 191, 315, 445, 481
Zinc Bloom 41
Zinc Spar 280
Zincite *15*, 197
Zinnwaldite 311
Zippeite 133
Zircon *15*, 355, 521
Zircon-favas 473
Zirconia 473
Zoisite *169*, 293

General Index

Locations of Minerals

*Figures indicate numbers of minerals found in the given locality;
bold figures refer to localities of illustrated minerals.*

Europe

Austria

Ankogel, Salzburg 572
Bleiberg, Carinthia 41, 55, 79, 172, 280, **484**
Eisenerz, Styria **63**
Greiner in Zemmtal, Tyrol 164, 395
Greiner Wald, Upper Austria 331
Gulsen near Kraubath, Styria 256, **457**
Habachtal, Salzburg 306
Hallein near Salzburg 44, **190**
Hallstatt, Salzkammergut 30, 44, 132, **224**
Hüttenberg, Carinthia 60, **96**, 383
Knappenwand in Untersulz-bachtal, Salzburg 290, 295
Kogel near Brixlegg, Tyrol **446**
Kraubath, Styria 65
Krieglach, Styria 239
Leoben, Styria 276
Leogang, Salzburg 56
Lisenz Alps, Sellrain, Tyrol **426**
Lölling near Hüttenberg, Carinthia 10, 407
Mariazell, Styria **542**
Mitterberg, Salzburg 15
Modriach, Styria **541**
Obergurgl, Ötztal, Tyrol 219
Oberpinzgau, Salzburg 339
Obirberg near Kappel, Carinthia 326, 406
Ötztal, Tyrol **285**
Pfunders, Tyrol 104, 162
Prägraten-on-Isel, East Tyrol 198, 348
Rauris, Pinzgau, Salzburg 293
Rothenkopf, Zillertal, Tyrol **104, 286**
Saualpe, Carinthia 293
Schellgaden, Salzburg 157
Schladming, Styria **180**
Schmirn near Matrei, Tyrol 100
Schwarzenbach near Bleiburg, Carinthia 402
Schwaz, Tyrol **250**
Sonnblick, Salzburg 119, **235**
Spittal-on-Drau, Carinthia 108
Sticklberg, Lower Austria **240**
Sulzbachtal, Oberpinzgau, Salzburg 162, 339
Teufelsmühle, Habachtal, Salzburg **339**
Trieben, Styria **405**
Veitsch, Styria 405
Weisseck, Salzburg **233**
Werfen, Salzburg **85**, 152, **239**
Zillertal, Tyrol 20, 248, 295, 359, **395**

Belgium

Berneau near Liège 169
Vielsalm, Ardennes 176, 275

Czechoslovakia

Banská Štiavnica, Slovakia **59**, 107, 142, 151, 186, 193, 218
Bělá, Bohemia **562**
Besednice, Bohemia **551**
Blatno, Bohemia **201**
Březová, Bohemia **464**
Cerhovice, Bohemia **68**
Chabičov, Moravia **314**
Chuchle, Bohemia 18
Cínovec, Bohemia 168, **171**, 207, 267, **311**, **323**, **455**
Čermíky near Kadaň, Bohemia 29
Červenica, Slovakia 202
Česká Lípa, Bohemia **88**
Dědova Hora, Bohemia **234**
Dobšiná, Slovakia **576**
Dolní Bory, Moravia **244**
Donovaly, Slovakia **437**
Doubice, Bohemia **90**
Dubník, Slovakia **139**, **535**
Duchcov, Bohemia **31**
Handlová, Slovakia **566**
Hauenštejn, Bohemia **84**
Hazlov, Bohemia **352**
Horní Blatná, Bohemia 425
Heřmanov, Moravia **345**
Hořenec, Bohemia **61**
Hrbek, Bohemia **148**
Hrubšice, Moravia 24, **298**
Jáchymov, Bohemia **38**, **48**, **133**, **137**, **185**, **249**, 251, 365, **373**, 443, **565**
Jílové, Bohemia **120**
Kadaň, Bohemia **247**
Karlovy Vary (Carlsbad), Bohemia 62, 320, **485**
Kaznějov, Bohemia **252**
Kladno, Bohemia 50, **123**
Komořany, Bohemia **127**
Kopanina, Bohemia **564**
Kozákov, Bohemia 96, **559**
Kralupy, Bohemia 18
Kraslice, Bohemia **354**, **574**
Kremnice, Slovakia 45
Krkonoše (Giant Mountains), Bohemia **553**
Křepice, Bohemia **121**
Krupka, Bohemia **4**, **362**, **444**
Kunštát, Moravia **569**
Kutná Hora, Bohemia 193, 448
Kuzmice, Slovakia 19
Levín, Bohemia **507**, **538**
Lovinobaňa, Slovakia **399**
Lubětová, Slovakia 271, **275**, **277**

Lukov, Bohemia **467**, **468**
Magurka, Slovakia 367
Maršíkov, Moravia 293, 304
Meclov, Bohemia **305**
Měděnec, Bohemia **510**
Mlýnky, Slovakia **60**
Mořina, Bohemia **170**
Nučice, Bohemia **169**, **404**
Obří důl, Giant Mountains, Bohemia **157**
Oloví, Bohemia **325**, **401**
Paškopole, Bohemia 467
Peřimov, Bohemia **111**
Podbořany, Bohemia 21
Podmoklice, Bohemia **528**
Podsedice, Bohemia **527**
Prešov, Slovakia **32**
Příbram, Bohemia **6**, **8**, 143, **145**, **175**, **316**, **321**, **327**, **335**, **374**, **375**, **378**, **379**, **381**, 443, **448**, **460**, **462**, **477**
Ratibořice, Bohemia **110**
Rožná, Moravia **208**, 211, **245**
Řepčice, Bohemia **77**
Sedlec near Carlsbad, Bohemia 21
Slavkov, Bohemia **163**, **214**, **258**, 261, **281**, **438**, **439**, **454**
Sliač, Slovakia **567**
Smrkovec, Bohemia **261**
Stříbro, Bohemia 153, **487**
Stupava, Moravia **575**
Špania Dolina, Slovakia 45, 228, **232**, **257**, **263**
Šternberk, Moravia **400**
Tajov near Banská Bystrica, Slovakia 131
Tatobity, Bohemia **557**
Tri Vody, Slovakia **422**
Ústí-on-Elbe, Bohemia **199**
Valchov, Moravia **141**
Valeč, Bohemia **95**
Vinařická hora, Bohemia **75**
Vrchlabí, Bohemia **563**, **570**
Zaječov, Bohemia **499**
Zbirov, Bohemia **160**
Železník, Slovakia **69**, **482**

Denmark

Faroe Islands, Nólsoy (Skútin Cave) **22**, **78**, **84**, **91**, **202**, 213

Federal Republic of Germany

Altenberg near Aachen **47**, **79**, 280, 283, 411
Arnsberg, Westphalia 367, 374
Aschaffenburg, Spessart 358, 458
Auerbach a. d. Bergstrasse 351, 433

List of Reference Books

Anderson, B. W.: Gem Testing. Heywood & Comp., Ltd, London 1947

Bloss, F. Donald: An Introduction to the Methods of Optical Crystallography. Holt, Rinehart & Winston, New York 1961

Bunn, Charles: Crystals — Their Role in Nature and in Science. Academic Press, New York — London 1964

Dake, H. C., Fleener Frank L., Wilson Ben Hur: Quartz Family Minerals. A Handbook for the Mineral Collector. Whittlesey House, McGraw-Hill, New York — London 1938

Dana, E. S.: A Textbook of Mineralogy. John Wiley & Sons, New York 1932

Dana, L., Hurlbut S.: Manual of Mineralogy. Wiley, New York 1952

Deer, W. A., Howie, R. A., Zussman, J.: Rock Forming Minerals. 1962—1964

Kraus, E. H., Hunt, W. F., Ramsdell, L. S.: Mineralogy — An Introduction to the Study of Minerals and Crystals. McGraw-Hill, New York — London 1936

Kraus, E. H.: Tables for the Determination of Minerals. McGraw-Hill, New York — London 1930

Kraus, E. H., Slawson, C. B.: Gems and Gem Materials. McGraw-Hill, New York — London 1941

Lindgren, W.: Mineral Deposits. McGraw-Hill, New York — London 1933

McCarthy, J. R.: Fire in the Earth — The Story of the Diamond. Robert Hale Ltd, London 1946

Phillips, F. C.: An Introduction to Crystallography, Longmans, Green & Co., London, New York, Toronto 1949

Read, H. H.: Elements of Mineralogy. Woodbridge Press, Ltd, Guilford 1939

Rogers, A. F., Kerr, P. F.: Optical Mineralogy. McGraw-Hill, New York — London 1942

Rogers, A. F.: Introduction to the Study of Minerals. McGraw-Hill, New York — London 1937

Taylor, H. E.: Wonders of the Earth's Crust. Sir Isaac Pitman & Sons, Ltd, London 1932

Wahlstrom, E. E.: Optical Crystallography. John Wiley & Sons, Inc., New York — London 1960

Webster, Robert: The Gemmologists Compendium. N.A.G. Press, Ltd, London 1947

Webster, Robert: Gems. Their Sources, Descriptions and Identification. London 1962

Winchel, A. N., Winchel, H.: Elements of Optical Mineralogy. John Wiley & Sons, Inc., New York — London 1951